SpringerBriefs in Applied Sciences and Technology

SpringerBriefs present concise summaries of cutting-edge research and practical applications across a wide spectrum of fields. Featuring compact volumes of 50 to 125 pages, the series covers a range of content from professional to academic.

Typical publications can be:

- A timely report of state-of-the art methods
- An introduction to or a manual for the application of mathematical or computer techniques
- A bridge between new research results, as published in journal articles
- A snapshot of a hot or emerging topic
- An in-depth case study
- A presentation of core concepts that students must understand in order to make independent contributions

SpringerBriefs are characterized by fast, global electronic dissemination, standard publishing contracts, standardized manuscript preparation and formatting guidelines, and expedited production schedules.

On the one hand, **SpringerBriefs in Applied Sciences and Technology** are devoted to the publication of fundamentals and applications within the different classical engineering disciplines as well as in interdisciplinary fields that recently emerged between these areas. On the other hand, as the boundary separating fundamental research and applied technology is more and more dissolving, this series is particularly open to trans-disciplinary topics between fundamental science and engineering.

Indexed by EI-Compendex, SCOPUS and Springerlink.

More information about this series at http://www.springer.com/series/8884

Richard Dvorsky · Ladislav Svoboda · Jiří Bednář

Nanoparticles' Preparation, Properties, Interactions and Self-Organization

EUROPEAN UNION
European Structural and Investment Funds
Operational Programme Research,
Development and Education

MINISTRY OF EDUCATION,
YOUTH AND SPORTS

Richard Dvorsky ⓘD
The Centre for Energy and Environmental
Technologies—Nanotechnology Centre,
IT4Innovations National Supercomputing
Center
VSB—Technical University of Ostrava
Ostrava Poruba, Czech Republic

Ladislav Svoboda ⓘD
The Centre for Energy and Environmental
Technologies—Nanotechnology Centre,
IT4Innovations National Supercomputing
Center
VSB—Technical University of Ostrava
Ostrava Poruba, Czech Republic

Jiří Bednář
The Centre for Energy and Environmental
Technologies—Nanotechnology Centre,
IT4Innovations National Supercomputing
Center
VSB—Technical University of Ostrava
Ostrava Poruba, Czech Republic

ISSN 2191-530X   ISSN 2191-5318   (electronic)
SpringerBriefs in Applied Sciences and Technology
ISBN 978-3-030-89143-5   ISBN 978-3-030-89144-2   (eBook)
https://doi.org/10.1007/978-3-030-89144-2

This Springer imprint is published by the registered company Springer Nature Switzerland AG
The registered company address is: Gewerbestrasse 11, 6330 Cham, Switzerland

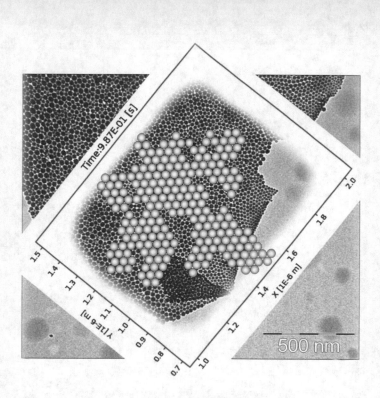

Time:9.87E-01 [s]

500 nm

Here's to the crazy ones.
The misfits.
The rebels.
The troublemakers.
The round pegs in the square holes.
The ones who see things differently.
They're not fond of rules.
And they have no respect for the status quo.
You can quote them, disagree with them,
glorify or vilify them.
About the only thing you can't do is ignore
them.
Because they change things.
They push the human race forward.
And while some may see them as the crazy
ones, we see genius.
Because the people who are crazy enough to
think they can change the world,
are the ones who do.
Steve Jobs
Their h-index is usually zero.

Preface

One of the fundamental concepts of modern nanoscience is the NANOPARTICLE, and it is also the name of the oldest used nanostructure in history. The first nanoparticles appeared in ancient Egypt in the fifth to fourth century BC in the form of "soluble gold" in the dyeing of pottery or in the production of ruby glass. Also very famous are the so-called Lycurgus cups, which were probably made in Alexandria at the beginning of the fourth century AD. Their glass exhibits a dichroic effect due to local plasmon resonance on finely dispersed gold nanoparticles. Nanoparticles of copper and silver metal with a diameter of 5–100 nm were identified in Renaissance glazes, which gave them an unmistakable shine. In 1857 Michael Faraday obtained colloidal gold particles by reduction of a tetrachlorosilane solution, and shortly afterwards colloidal particles of other metals were prepared. This represents the roots of a new direction called colloid chemistry, which has now become as integral part of nanoscience as the physics and chemistry of nanoparticles. In 1925, Nobel laureate Richard Adolf Zsigmondy was one of the first scientists to visualize nanoparticles in gold salts using an optical ultramicroscope, and probably the first to use the unit of one nanometer to characterize their size. The most widespread type of "new-age" nanoparticles has long been the fine fraction of 10–500 nm carbon particles that are bound into larger soot aggregates. These have been the most widely used nanomaterial for the last 100 years, with an estimated current annual production of 6 million tons. An oft-cited moment in the development of nanoscience is Nobel laureate Richard Feynman's famous lecture called "There's Plenty of Room at the Bottom" given at the 1959 meeting of the American Physical Society at Caltech. In it, he literally challenged the scientific community to "conquer the nanoworld", and he followed his speech with the question "Why can't we put the 24 volumes of the Encyclopaedia Brittanica on a pinhead?" and the curious announcement of a reward for writing one page of a regular book on 1/25,000 of the original area, among others. This episode became legendary thanks to Richard Feynman's famous scientific reputation, but the real start of nanoscience came a little later. At a conference in 1974, Norio Taniguchi of the University of Tokyo was the first to use the term "nanotechnology", and independently, Eric Drexler used the term again in his first paper, which popularized the concept of nanotechnology and also defined the direction of molecular nanotechnology. In the 1980s, Eric Drexler promoted the technological possibility of manipulating individual atoms to form functional nanostructures, such as nanomotors and similar mechanisms that assemble individual atoms into purposeful structures. He has published his views in two influential books. The first of these is entitled *Molecular Engineering: An Approach to Development of General Capabilities for Molecular Manipulation*, and was published in 1981 in the Proceedings of the National Academy of Sciences of the United States of America. He described the actual methods of assembling atoms into useful functional structures in 1986 in his second book, *Engines of Creation: The Coming Era of Nanotechnology*. In this legendary treatise, he also comes up not only with the idea of nanobots, but also of self-replicating automata, for which further spontaneous evolution of structure is not excluded. In 1986, Eric Drexler founded the Foresight Institute with the mission of "Preparing for Nanotechnology", which is still developing key areas such as molecular machines, bionanotechnology, computational nanostructures and more. One of the landmarks of nanotechnology in the field of nanoparticles was the 1985 discovery of a third allotropic form of carbon, fullerene, for which Harry Kroto, Richard Smalley and Robert Curl won the Nobel Prize in 1996. In the 1990s, Wolfgang Krätschmer and Donald Huffman discovered a method for the massive industrial synthesis of fullerenes, thus creating the conditions for wider research into their properties, especially in the field of their functionalization with other active dopants and whole complexes. As a result of the growing application potential of nanotechnology, the United States adopted the "National Nanotechnology Initiative" in 2000 to support and develop the four main objectives in the area. This is "Advance a world-class nanotechnology research and development program; Foster the transfer of new technologies into products for commercial and public benefit; Develop and sustain educational resources, a skilled workforce, and a dynamic infrastructure and toolset to advance nanotechnology; and Support responsible development of nanotechnology". According to the Statistics Portal for Market Data "Statista", the global market value of nanotechnology will reach approximately USD 76 billion in 2020 and is showing a steady growth rate.

Nanoscience deals with basic research in the field of nanoscale dimensions. Its multidisciplinary nature is based on the experimental experience that the properties of a substance at the "nano" scale are in many cases not predictable from its investigation at larger macroscales. Significant changes are caused not only by the modification of physicochemical parameters due to crystallinity or the increase in the fraction of surface atoms in nanoparticles, but also by hybrid behavior, which gradually includes various quantum effects.

The motivation for writing this book is the intention to describe one aspect of the physical interaction between nanoparticles in more detail. It involves the collective action of Van der Waals forces in the self-organization of large ensembles of nanoparticles into higher structures after releasing their fixation in a frozen liquid dispersion at a vacuum sublimation interface. Based on systematic research into the above phenomenon, a basic version of the "controlled sublimation technology" was developed and the fact that it has been granted patent protection in several countries confirms both the novelty of the technical solution and the high relevance of detailed research into the relevant physical processes. The application potential of controlled sublimation technology is increasing due to the current requirements for the preparation of materials with high specific surface area, such as in the case of efficient sorbents, catalysts and recently especially photocatalysts.

Parasitic anthropogenic production of nanoparticles in the form of exhalations has been occurring in industrial areas for over a century. However, more massive emissions to air and water have occurred especially in the last few decades as a result of the explosive growth of the automotive industry and as a result of conscious human activity in the implementation of nanotechnology in industry. Like any new technology, nanotechnology poses risks in addition to its undeniable benefits in terms of its negative impact on the overall environment and the human body. In addition to these parasitic emissions, a completely new moral hazard has recently emerged in the area of irresponsible gene manipulation. Here, as nowhere else, the relevant field of almost Faustian research is in danger of becoming, in the words of the alchemists, "a science without a conscience". Without legislative constraints that clearly define the risks and the relevant safety directives, society cannot successfully develop. Given that current knowledge in the field of nanoscience is by no means sufficient, the scientific information necessary for the above-mentioned legislative restrictions does not exist in many areas. In this situation, it is very important to call on nanotechnology companies in particular to observe the 'precautionary principle' in their activities, so that they take economically and socially viable precautions against the emissions in question, even if harmful effects have not yet been clearly demonstrated. This transitional situation will gradually be transformed into firm legislative rules in line with the development of knowledge in nanoscience. Working groups within both the European Commission and the OECD are preparing a scientifically sound definition of nanomaterials and based on this definition, a methodology for their classification. Globally, nanotechnology risks are addressed by organizations such as EFSA (European Food Safety Authority), SCENIHR (Scientific Committee on Emerging and Newly Identified Health Risks), US-EPA (Environmental Protection Agency of the USA), OECD (Organization for Economic Co-operation and Development), FSA (Food Standard Agency) and many others. These organizations issue recommendations and draft documents on the safety of nanomaterials, such as the Nanotechnology White Paper (US-EPA), which specifies the potential negative effects of selected nanomaterials. An illustrative example is the well-known and widely used nanomaterial TiO_2, which has found applications in many different industries since the 1920s. However, it was not until more than almost 90 years later that titanium dioxide dust was classified as IARC Group 2B carcinogen when inhaled by the International Agency for Research on Cancer (IARC) in 2006, meaning that it is probably carcinogenic to humans. According to a 2014 report by the Scientific Committee on Consumer Safety, the behavior of TiO_2 nanoparticles has been described as genotoxic, carcinogenic and photosensitizing. The delegated regulation classifies TiO_2 as a category 2 suspected carcinogen by inhalation under EU Regulation (EC) No 1272/2008. The classification was followed by an opinion published by the Risk Assessment Committee (RAC) of the European Chemicals Agency (ECHA) in 2017. The regulation entered into force on 9th September 2021. This means that products containing TiO_2 must carry a cancer warning on the label, which applies to mixtures in powder form containing 1% or more TiO_2 with an aerodynamic diameter of 10 µm or less. For other forms and mixtures, the classification proposes specific notes to inform users of precautions to be taken to minimize the hazard. While liquid and some solid mixtures are not classified, those containing more than 1% TiO_2 should be subject to special warning phrases and labelling. Despite the availability of the aforementioned research and reports, the scientific evaluation of TiO_2 carried out by the European Food Safety Agency (EFSA) in 2018 confirmed its safety as a food additive (E171).

Ostrava Poruba, Czech Republic

Richard Dvorsky
Ladislav Svoboda
Jiří Bednář

Acknowledgements

We acknowledge support from the project financed by the project Development and Education, financed by Operation programe Research of EUROPEAN STRUCTURAL AND INVESTMENT FUNDS, by the project Gamma PP1 (TP01010036) by Technology Agency of the Czech Republic, project DGS/TEAM/2020-001 "Organic and inorganic pathogenic nanoparticles and the formation of appropriate protective barriers based on electroactive nanomaterials" funded by Operational Programme Research, Development and Education (CZ.02.2.69/0.0/0.0/19_073/0016945), Operational Programme Research, Development and Education financed by the EU and from the budget of the Czech Republic "Innovative therapeutic methods of musculoskeletal system in accident surgery" (CZ.02.1.01/0.0/0.0/17_049/0008441), European Regional Development Fund in the IT4Innovations national supercomputing center—path to exascale project (EF16_013/0001791). Especial thanks take to J. Nynka for the overall support of the book writing.

Contents

Symbols

A	Phenomenological proportionality constant (1.1.2.1)
a	Lattice constant (1.1.2.1)
A	Reactant A (1.1.2.3)
a_B	Bohr radius (1.1.3.2)
a_{ex}	Bohr radius of the exciton (1.1.3.2)
A_{ij}	Hamaker constant (4.1.1)
$A_m\,B_n$	Reaction product (1.1.2.3)
B	Reactant B (1.1.2.3)
B	External magnetic induction (1.1.3.3)
c	Dispersion concentration (1.2.4.3)
c	Concentration (1.1.2.2)
C_{C60}	Percentage fraction of c_{sat} (5.1.1)
c_{sat}	Saturated concentration (1.2.4.3)
C_{dis}	Volume fraction size of microdroplets (5.1.1)
C_N	Newton's resistive coeficient (3.2.4)
C_{nC60}	Fullerite nC_{60} nanoparticle concentration (4.1.2)
c_o	Volume concentration (5.1.1)
c_s	Residual concentration of methylene blue after sorption (5.2.3.3)
c_{sat}	Saturated concentration of fullerene in toluene (5.1.1)
c_{sat}	Saturation concentration (1.2.4.3)
d	Diameter (1.1.2.3)
$\bar{d}$	Mean diameter (5.2.2)
D	Electric induction (1.1.3.1)
D	Particle diameter (1.1.3.3)
d	Minimal distance between particle surfaces (4.1.1)
d_{at}	Characteristic diameter of one atom in the particle (1.1.1)
D_o	Critical diameter between single-domain and multidomain states of nanoparticle (1.1.3.3)
dF	Helmholtz energy differential (1.1.2.2)
d_i	i-category of particle size diameter (3.3.1)
d_{min}	Minimum distance between particle surfaces (4.1.2)

dn^s	Substance amount in solid differential (1.1.2.2)
dn^{sl}	Substance amount in in solution differential (1.1.2.2)
D_{sp}	Particle diamater corresponding to V_{sp} (1.1.3.3)
E	Mean energy of harmonic oscillators at the melting temperature T_m (1.1.2.1)
$\mathbf{E}$	Electric intensity (1.1.3.1)
e	Electron charge (1.1.3.1)
E	Wannier Hamiltonian eigenvalue (1.1.3.2)
E_0	Electric field amplitude (1.1.3.1)
$\mathbf{e}_1$	Base unit vector for coordinate x j (1.1.3.1)
E_b	Energy barrier separating two antiparallel magnetizations (1.1.3.3)
E_{ex}	Exciton ground energy in potential well (1.1.3.2)
$E_g{}^{bulk}$	Bulk band gap (1.1.3.2)
$E_g{}^{nano}$	Nanoparticle band gap (1.1.3.2)
E_k	Effective kinetic energy of the particle (3.2.3)
$E_k{}^{eqp}$	Mean kinetic energy in a liquid (3.2.3)
E_m	Energy contribution to antiparallel magnetization change (1.1.3.3)
E_{max}	Maximum magnetic polarization energy (1.1.3.3)
E_{min}	Minimum magnetic polarization energy (1.1.3.3)
$\mathbf{E}_{out}$	Electric intensity outside the nanoparticle (1.1.3.1)
$E_{Ry}{}^{ex}$	Rydberg exciton energy (1.1.3.2)
$\mathbf{F}$	Total force acting on the electron (1.1.3.1)
$\mathbf{F}_0$	Force acting on individual electrons (1.1.3.1)
F_N	Newton's resistive force acting on a particle (3.2.4)
F_N	Newton's resistive force (3.2.4)
$\mathbf{F}_P$	Polarization force (1.1.3.1)
G	Gravitational force acting on a particle (3.2.4)
g	Gravitational acceleration (3.2.4)
G_1	Gibbs energy of initial state (1.1.2.3)
G_2	Gibbs energy of final state (1.1.2.3)
G_{bulk}	Gibbs energy of bulk (1.1.2.3)
G_{in}	Input Gibbs energy (1.1.2.3)
G_{out}	Output Gibbs energy (1.1.2.3)
G_s	Gibbs energy (1.1.2.3)
h	Thickness of the surface atom monolayer (1.1.2.1)
h	Planck constant (1.1.2.3)
H	Intensity of the external field (1.1.3.3)
h	Transition layer thickness (3.2.4)
$\hbar$	Reduced Planck constant (1.1.3.2)
H_c	Coercive magnetic intensity (1.1.3.3)
H_{c0}	Coercive magnetic intensity for a bulk (1.1.3.3)
H_w	Wannier Hamiltonian (1.1.3.2)
i	Imaginary unit (1.1.3.1)
I_{in}	Inside flow of material (1.1.2.2)
I_{out}	Outside flow of material (1.1.2.2)

J_{sub}	Flow of the sublimation wind (3.2.4)
k	proporcionality parameter (1.1.2)
k	Reaction rate coefficient (1.1.2.3)
k	Multiplication parameter of vacuum reduction under saturated vapor pressure (3.2.3)
k	Reaction rate constant (5.2.3.2)
k_1	Microscopic reaction rate constant (1.1.2.3)
K_a	Anisotropy coefficient (1.1.3.3)
k_B	Boltzman constant (1.1.2.1)
λ	Dimensionless parameter $p/p_{sat} \leq 1$ (3.2.4)
$\mathbf{m}$	Magnetic moment of the particle (1.1.3.3)
M	Magnetization of the particle (1.1.3.3)
$M(r)$	Particle mass (1.1.2.3)
m^*	Effective mass (3.2.3)
m°_e	Effective electron mass (1.1.3.2)
m°_h	Effective hole mass (1.1.3.2)
m_e	Electron mass (1.1.3.1)
M_{ex}	Reduced mass of the exciton (1.1.3.2)
M_{H2O}	Molar mass of water (3.2.3)
MIC	Minimum inhibitory concentration (5)
MIC_0Ag	Minimum inhibitory concentrations for Ag in pure form (5.2.3.5)
MIC_0Zn	Minimum inhibitory concentrations for Zn in pure form (5.2.3.5)
MIC_{Ag}	Minimum inhibitory concentrations for Ag in composite form (5.2.3.5)
MIC_{Zn}	Minimum inhibitory concentrations for Zn in composite form (5.2.3.5)
m_{nC60}	Mass of nC_{60} nanoparticles (4.1.2)
M_s	Spontaneous magnetization of the nanoparticle (1.1.3.3)
N_A	Avogadro constant (1.1.2.3)
N_{at}	Number of atoms in the nanoparticle (1.1.1)
$N_{at}(R)$	Number of atoms in globular particle with radius R[m] (1.1.1)
$N_{at}(S)$	Number of atoms in the s-shell cluster (1.1.1)
n_e	Mean electron density (1.1.3.1)
n_i	Refractive index (4.1.1)
n_s/n	Surface atoms fraction (1.1.2.1)
n_v/n	Internal atoms fraction (1.1.2.1)
$\mathbf{P}$	Electric polarization (1.1.3.1)
p	Momentum (1.1.3.1)
p	Pressure in the vacuum recipient (3.2.3)
$P(d)$	Statistical distribution frequency function (4.2.2)
P_h	Vapor pressure at a very small height above the sublimation interface (3.2.4)
p_0	Absolute pressure of the liquid (1.2.3.4)
p_{sat}	Saturated vapor pressure of water (3.2.3)
$P_v(d)$	Statistical distribution of total volume (3.3.1)
p_{vac}	Partial vapor pressure of water (3.2.3)
q_s	Sorption capacity (5.2.3.2)

R	Radius of globular particle (1.1.1)
r	Particle radius (1.1.2)
R	Universal gas constant (1.1.2.2)
$\mathbf{r}$	Position radius vector (1.1.3.1)
r	Distance from nanoparticle center (1.1.3.1)
$R(r)$	Bessel functions (1.1.3.2)
$\overline{\rho}_o$	Average ice density (3.2.3)
R_0	Microdroplet radius (5.1.1)
R_1	First particle radius (4.1.1)
R_2	Second particle radius (4.1.1)
R_{bmax}	Initial radius of the collapsing cavitation bubble (1.2.3.4)
r_{bulk}	Characteristic radius of the macroparticle (1.1.2.1)
RE	Regeneration efficiency (5.2.3.2)
r_i	Particle radii; $i = 0, 1, 2$ (1.1.2.2)
r_{nano}	Characteristic radius of the nanoparticle (1.1.2.1)
R_{oo}	Minimum radius of the emulsion droplet (5.1.1)
R_{par}	Radius of spherical particle (5.1.1)
s	Serial number of the atomic shell in the cluster (1.1.1)
S	Area of ice (3.2.3)
S	Cross-section of particle (3.2.4)
$S(r)$	Particle surface (1.1.2.3)
S_r	Surface of spherical nanoparticle (1.1.2.2)
S_{sub}	Free sublimation surface area (3.2.3)
T	Temperature (1.1.2.2)
t	Time (1.1.3.1)
t	Time of sublimation (3.2.3)
T_b	Blocking temperature between the ferromagnetic and blocked state of the particle (1.1.3.3)
T_m	Melting temperature (1.1.2.1)
T_{m0}	Melting temperature for bulk material (1.1.2.1)
$U(d)$	Hamaker interaction potential of two same particles (4.1.2)
$U_{12}(d)$	Hamaker's interaction potential of two spherical particles (4.1.1)
υ	Reaction rate (1.1.2.3)
V	Potential (1.1.3.2)
V	Particle volume (1.1.3.3)
$\overline{\upsilon}^2$	Mean square velocity of water molecules emitted from the sublimation interface (3.2.4)
$\overline{\upsilon}_\perp^2$	Quadratic velocities perpendicular to the sublimation interface (3.2.4)
υ_{at}	Elementary cubic equivalent of one atom in the particle (1.1.1)
V_m^s	Molar volume in the solid state (1.1.2.1)
υ_o	Retraction speed of the sublimation interface (3.2.3)
$\upsilon_{o\,exp}$	Experimental retraction speed of the sublimation interface (3.2.3)
V_{part}	Volume of globular particle with radius R [m] (1.1.1)
υ_{rms}	Instantaneous velocitiy of globular silicon microparticles (3.2.3)
V_s	Volume of solid part (1.1.2.2)

V_{sp}	Limit volume of the particle with blocking temperature T_b (1.1.3.3)
υ_{Zn}	Zinc mass fraction (5.2.3.5)
x	Time-dependent oscillation deviation (1.1.3.1)
x	Distance from the sublimation interface to the "practically infinite" volume of the vacuum recipient (3.2.4)
x_0	Equilibrium point of undamped linear oscillations (1.1.3.1)
$x_{nC_{60}}x$	Fullerite nC_{60} nanoparticle mass fraction (4.1.2)
$Y_l^m(\theta, \varphi)$	Spherical harmonics (1.1.3.2)
α	Proportionality coefficient (5.1.1)
γ	Surface energy (1.1.2.3)
γ	Areal energy density of the domain wall (1.1.3.3)
δ	Characteristic relative fluctuation of particle physical parameters (1.1.1)
δ^E	Relative change in energy (1.1.3.2)
Δ^E	Energy barrier between two polarizations (1.1.3.3)
$\Delta^E_{\uparrow\downarrow}$	Difference of the potential energies of two antiparallel polarizations (1.1.3.3)
Δ^G	Change in the Gibbs energy (1.1.2.3)
$\Delta^H{}_m$	Molar enthalpy of melting (1.1.2.1)
Δ^m	Weight of ice (3.2.3)
Δ^n	Number of water molecules emitted into the vacuum (3.2.3)
Δ^P	Change in the momentum (1.1.3.1)
Δ^P	Pressure difference (3.2.4)
Δ^t	Time difference (1.1.3.1)
ε	Permittivity (1.1.3.2)
ε	Volume energy density (1.1.3.3)
ε_0	Vacuum permittivity (1.1.3.1)
ε_{in}	Permittivity inside the particle (1.1.3.1)
ε_{out}	Permittivity outside the particle (1.1.3.1)
ε_r	Relative permittivity (1.1.3.1)
ε_{ri}	Relative permittivity (4.1.1)
θ	Polarization energy orientation (1.1.3.3)
θ_{max}	Angle for maximum magnetic polarization energy (1.1.3.3)
θ_{min}	Angle for minimum magnetic polarization energy (1.1.3.3)
κ	Contact surface fraction (3.2.3)
λ	Wavelength of light (1.1.3.1)
μ^s	Chemical potential of the substance in the solid (1.1.2.2)
μ_{sl}	Chemical potential of the substance in the solution (1.1.2.2)
μ_0	Vacuum permeability (1.1.3.3)
μ_B	Bohr magneton (1.1.3.3)
ν_e	Absorption maximum in ultraviolet area (4.1.1)
ρ	Material density (1.1.2.3)
ρ_h	Density of flowing vapor in a layer of thickness h above the sublimation interface (3.2.4)
ρ_{H_2O}	Water density (4.1.2)
ρ_{nC60}	Density of crystalline fullerite (5.1.1)

ρ_{part} Particle density (3.2.4)
ρ_{part} Particle density (3.3.1)
ρ_{sat} Saturation vapor density (3.2.4)
ρ_{Si} Silicon density (3.2.3)
ρ_{sol} Particle density (5.1.1)
σ Mean vibration deflection (1.1.2)
$\Sigma(d)$ Surfaces related to a unit volume of 1 cm^{-3} (1.1.2.3)
$\Sigma(r)$ Specific surface (1.1.2.3)
$\Sigma_{mol}(r)$ One mole of solid particulate material (1.1.2.3)
σ_{sl} Interfacial tension (1.1.2.2)
σ_{bon} Critical vibration deflection (1.1.2)
σ_v Parameter of internal vibrations (1.1.2.1)
τ Mean time between two collisions (1.1.3.1)
τ Period of water molecules emitted into the vacuum (3.2.3)
τ_0 Time constant in Neel-Arrhenius equation (1.1.3.3)
τ_m Measurement time (1.1.3.3)
τ_N Neel's relaxation time (1.1.3.3)
$\Phi(t)$ Potential (1.1.3.1)
$\Phi(\mathbf{x})$ Exciton motion function (1.1.3.2)
$\Phi_{in}(t)$ Potential inside the particle (1.1.3.1)
Φ_{nC60} Releasing frequency of dispersed nC$_{60}$ nanoparticles from the unit area (4.1.2)
$\Phi_{out}(t)$ Potential outside the particle (1.1.3.1)
$\varphi_{rest}(\mathbf{x})$ Dimensional restriction function (1.1.3.2)
$\psi_{bloch}(\mathbf{x})$ Bloch function (1.1.3.2)
ω Frequency (1.1.3.1)
ω_0 Resonant frequency (1.1.3.1)
ω_p Plasma frequency (1.1.3.1)
$\overline{\rho}_o$ Average ice density (3.2.3)
$\overline{\upsilon}^2$ Mean square velocity of water molecules emitted from the sublimation interface (3.2.4)
$\overline{\upsilon}_{\perp}^2$ Quadratic velocities perpendicular to the sublimation interface (3.2.4)

List of Figures

List of Tables

Chapter 1
Nanoparticles—Their Specific Properties and Origin

1.1 Nanoparticles and Their Basic Properties

1.1.1 Terminology Limit of 100 nm and Its Reasoning

Nanoparticles can be considered, with a certain license, as "inhabitants of the disputed territory" between the classical and quantum world. Due to the small range of the statistical set of their atoms, the application of classical statistical thermodynamics is very problematic in the description, and the definition of some thermodynamic quantities is even not possible in these conditions (such as the temperature of the set of 60 carbon atoms in the fullerene nanoparticle C_{60}).

The very concept of "nanoscale" has its basis in practical experience with very small dimensions of material objects, which exhibit properties different from conventional macroscopic bodies of the same material. Considering that the dimensional "limits of different behavior" are individual for different properties, the following unifying categorization of the term "nanoscale" has been formally adopted by the International Organization for Standardization (ISO) [1]:

"Nanoscale: Size ranges from approx. 1 to 100 nm.

Note 1: Properties that are not extrapolation from a larger size will typically, but not exclusively, be manifested at this dimensional scale. For such properties, the size limits are considered approximate.

© The Author(s), under exclusive license to Springer Nature Switzerland AG 2022
R. Dvorsky et al., *Nanoparticles' Preparation, Properties, Interactions and Self-Organization*, SpringerBriefs in Applied Sciences and Technology,
https://doi.org/10.1007/978-3-030-89144-2_1

Note 2: The lower limit in this definition (approximately 1 nm) is introduced to prevent individual small groups of atoms or parts of nanostructures (which may result from the absence of a lower limit) from being labeled as nanoobjects."

In the case of objects with dimensions close to the lower limit of the "disputed territory," such as quantum dots, new quantum effects are already beginning to appear, such as a change in the width of the band gap of semiconductors. In total, however, it is still true that nanoscale is neither classical macro-objects nor objects of a purely quantum nature. One of the basic parameters that cause different behavior of nanoparticles is the fraction of atoms that create the surface of an object.

Table. 1.1 shows the tendency of the fraction of surface atoms to decrease as the number of occupied shells in the cluster increases and thus also its size. In this context, it is interesting to compare the dependence of the number of cluster atoms on its size according to the geometry of the volume occupied by individual atoms.

$$
\begin{aligned}
a) &\rightarrow \left\{ \begin{array}{c} v_{at} \approx d_{at}^3 \\ V_{part} = \frac{4}{3}\pi R^3 \end{array} \right\} \rightarrow N_{at}(R) \approx \frac{V_{part}}{v_{at}} = \frac{4}{3}\pi \left(\frac{R}{d_{at}}\right)^3 \\
b) &\rightarrow \left\{ \begin{array}{c} N_{at}(s) = \frac{10}{3}s^3 - 5s^2 + \frac{11}{3}s - 1 \\ R(s) \approx \left(s + \frac{1}{2}\right)d_{at} \end{array} \right\} \rightarrow \\
&\rightarrow N_{at}(R) \approx \frac{10}{3}\left(\frac{R}{d_{at}} - \frac{1}{2}\right)^3 - 5\left(\frac{R}{d_{at}} - \frac{1}{2}\right)^2 + \frac{11}{3}\left(\frac{R}{d_{at}} - \frac{1}{2}\right) - 1
\end{aligned}
\tag{1.1}
$$

The number of atoms $N_{at}(R)$, determined by the Eq. (1.1a), indicates the number of elementary cubic equivalents $v_{at} = d_{at}^3$, representing one atom at full coverage of the total cluster volume. The same quantity, expressed by the Eq. (1.1b), is the

Table 1.1 Dependence of the fraction of surface atoms cluster on their shell structure (figures of clusters taken from work [2])

	1 shell	2 shells	3 shells	4 shells	5 shells	6 shells
Number of atoms:	13	55	147	309	561	1415
Surface atoms:	92%	76%	63%	52%	45%	35%

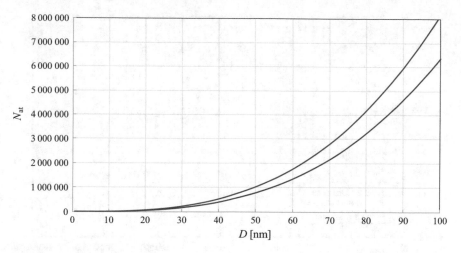

Fig. 1.1 Model dependence of the number of atoms ($d_{at} \approx 0.4$ nm) in a cluster on its diameter D determined on the basis of a Formula (1.1)a) and the same dependence determined on the basis of the Formula (1.1)b)

result of the gradual geometric layering of atoms into individual shells [3] Mackay icosahedron, which best approximates the initial cluster structure of the germs and the subsequent growth phase of the individual solid phase particles (Fig. 1.1).

The upper limit of the nanoscale thus corresponds to particles with characteristic numbers up to about 10^7 atoms. These limit systems already show relatively high relative fluctuations δ of physical parameters and do not meet the statistical conditions for a good definition of thermodynamic parameters (Fig. 1.2).

The relevant thermodynamic quantities cease to be realistic state variables in this situation. The characteristic value of relative fluctuations at the edge of the dimensional limit of the nanoscale 100 nm is approx. $3.5 \cdot 10^{-2}\% \gg 0$ and no longer meets the requirements of the central limit theorem of statistical thermodynamics [4].

$$\delta(N_{at}) \sim \frac{1}{\sqrt{N_{at}}} \cdots \lim_{N_{at} \to \infty} \delta(N_{at}) \to 0 \qquad (1.2)$$

If practical experimental devices allow to determine weights in the order of ng [5], then for these amounts of substances, e.g., 1 nmol of carbon, and the value of the characteristic relative fluctuation is δ (1 nmol C) $= 4.1 \cdot 10^{-8}\% \to 0$. Thermodynamic quantities for such macro-objects are already very well defined here and practically meet the condition of the central limit theorem of statistical thermodynamics.

The "NanoImpactNet Nomenclature Version 3" published in 2010 by The European Network on the Health and Environmental Impact of Nanomaterials is a useful resource for further guidance on terms relevant to area of nanoscale [6].

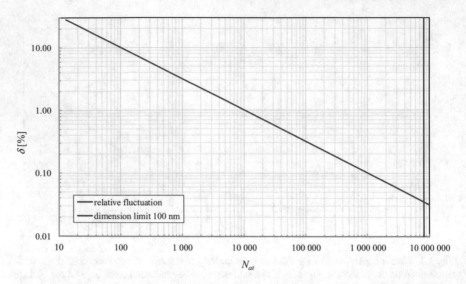

Fig. 1.2 Dependence of characteristic relative fluctuation of physical parameters $\delta \sim 1/\sqrt{N_{at}}$ on the number of nanoparticle atoms. Its value at the limit of the dimensional limit of 100 nm is about $3.5 \cdot 10^{-2}\%$

1.1.2   Binding Energy and Individual Stability
of Nanoparticles

The basic parameter of the binding interaction between molecules and atoms in macrovolumes ("bulk") of materials is the value of their binding energy; see Table. 1.2. This, together with the crystal structure, determines most of the important material parameters, such as melting point and hardness.

With the gradual reduction of the particle size, the different binding and thus also the vibrational conditions in the surface layer at the highly curved phase interface are an increasingly significant effect.

Table 1.2 Overview of binding energy values for basic types of bonds

Type of bond	Binding energy (J/mol)
Covalent	$(4–6) \times 10^5$
Metallic	$(2–4) \times 10^5$
Ionic	$(2–4) \times 10^5$
Hydrogen	$(2–3) \times 10^4$
Van der Waals force	$(4–8) \times 10^3$

1.1.2.1 Decrease in Melting Point for Small Particle Sizes

Closely related to this fact is also the reduced melting point of nanoparticles, which is a consequence of the morphology of small dimensions and energy conditions inside the nanoparticles. The relatively high proportion of surface atoms can be considered as dominant, and the binding energy of surface atoms with the surrounding atoms is significantly lower at a considerably curved surface than at the planar phase interface.

Based on the analysis of the binding stability of vibrating molecules, Lindemann proposed for the first time a simple criterion for the melting condition of a material [7]. If the mean vibration deflection σ exceeds the critical bond value σ_{vaz} (at least 10% of the distance and the nearest neighbors), a vibrating particle is released from the bond in the material:

$$\sigma \geq \sigma_{\text{vaz}} \approx \kappa \cdot a\kappa = 1.1 \tag{1.3}$$

Given the natural difference between the bonding structure in volume and on the surface of a particle of radius r, the mean standard deviation of vibrations σ can also be written as the sum of vibration contributions within σv and on surface σs, weighted by respective fractions of the numbers of internal (nv/n) and surface (ns/n) atoms [8, 9]

$$\sigma = \frac{n_v \sigma_v + n_s \sigma_s}{n_s + n_v} \tag{1.4}$$

In one of the first models, F.G. Shi introduced $\sigma(\lambda)$ as a function of the ratio $\lambda = n_s/n$ of the number of atoms on the surface and in volume.

$$\sigma(\lambda) = \frac{\sigma_v + \lambda \sigma_s}{\lambda + 1} \tag{1.5}$$

Assuming the same volume concentration of atoms inside and in the surface monolayer of atoms of thickness h, this applies

$$\lambda = \frac{n_s \sim 4\pi r^2 h}{n_v \sim \left(\frac{4}{3}\pi r^3 - 4\pi r^2 h\right)} = \frac{3h}{r - 3h} \tag{1.6}$$

For large-sized macroparticles, the curvature of the surface with respect to the dimensions of the atom is negligible, and the size of the radius can be considered as practically infinite $r \to \infty$ ($\lambda \to 0$).

$$\lim_{r \to \infty} \sigma(\lambda(r)) = \sigma(0) = \sigma_v. \tag{1.7}$$

The parameter of internal vibrations σv therefore has the meaning of the standard deviation $\sigma(0)$ of vibrations in infinite volume.

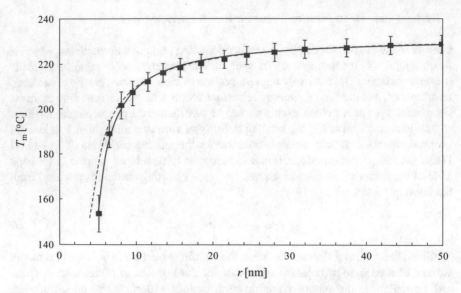

Fig. 1.3 Experimental dependence of the melting temperature of S_n nanoparticles on their radius r is represented by graph points and a red-dashed line presented in the work [10] is calculated on the basis of thermodynamic analysis of Hanszen's work [11]. The blue solid line represents the resulting fit based on the new theoretical derivation in this work (below) and coincides significantly better with the experimental data

In this work [10], the temperature dependence of the melting point T_m of tin was measured by the microcaloric method and its final course is represented by experimental points in Fig. 1.3.

Below it is presented a very simple new derivation of the dependence of the melting temperature on the radius r of a nanoparticle based on the Lindeman condition (1.3) and a model of a set of its atoms as harmonic oscillators with a mean energy $E = kT_m$ at the melting temperature T_m. Based on the mechanical formulation, the mean energy of simple linear oscillators in such a set is proportional to the mean value of the square of the critical bonding vibration deviations $E \sim \sigma_{vaz}{}^2$ and therefore holds

$$T_m \sim \sigma_{vaz}^2. \tag{1.8}$$

Unlike the model of Shi [8], presented work starts directly from the dependence of the critical bonding deviation $\sigma_{vaz}(\lambda)$ on the ratio $\lambda = n_s/n_v$. Based on experimental experience, it is clear that the decrease of the critical bonding differential deviation $d\sigma_{vaz}(\lambda)$ is proportional to its current size $\sigma_{vaz}(\lambda)$ and the increase of the ratio $d\lambda$

$$d\sigma_{vaz}(\lambda) \sim -\sigma_{vaz}(\lambda)d\lambda \rightarrow d\sigma_{vaz}(\lambda) = -A \cdot \sigma_{vaz}(\lambda)d\lambda. \tag{1.9}$$

By integrating the differential Eq. (1.9) within limits $(0, \lambda) \leftrightarrow (r_{bulk} \to \infty, r_{nano} < < \infty)$ for the phenomenological proportionality constant A, we obtain after substitution (1.6) specific mathematical form of the dependence of the critical bonding deviation $\sigma_{vaz}(r)$ on the radius r

$$\sigma_{vaz}(r) = \sigma_{vaz}(\infty)e^{-\frac{3hA}{r-3h}}. \tag{1.10}$$

After substitution of the model relationship (1.8), mathematical form (1.10) turns to the final formula of the dependence of the melting temperature $T_m(r)$ on the radius r

$$T_m(r) = T_{m0}e^{-\frac{6hA}{r-3h}}. \tag{1.11}$$

Fit of this phenomenological formula to experimental data of work [10] on Fig. 1.3 (blue line) shows a significantly better agreement than in the original work, especially in the area of very small radii.

In contrast to the above microstructural descriptions, the mentioned effect is currently also successfully described on the basis of a purely thermodynamic theory, assuming the continuity of the course of the chemical potential at the phase interface sol.-liq. [12] and [13]. Parameters determining the melting temperature T_m in the Gibbs–Thomson Eq. (1.12) [14]

$$T_m(r) = T_{m0}\left(1 - \frac{2\sigma_{sl}V_m^s}{\Delta H_m r}\right), \tag{1.12}$$

correspond gradually to the interfacial tension σ_{sl}, the molar volume in the solid state V_m^s, the molar enthalpy of melting ΔH_m and the radius r of the particle. An overview of its application to melts of indium, silver, palladium, copper and cadmium at work [12] is in good agreement with experimental data.

1.1.2.2 Ostwald Ripening

Equation (1.12) from the previous paragraph describes the so-called Gibbs–Thompson effect of curvature of the surface of small particles, which is also one of the main causes of increasing the equilibrium saturated concentration over the curved surface. The bond of atoms is significantly lower here than at the strictly planar interface (at $r \to \infty$). The equilibrium state of a particle of a given material in a solution of the same material in an inert solvent is due to the equality of flows of material outside I_{out} and inside I_{in} the particle by surface of phase interface $I_{out} + I_{in} = 0$. This corresponds to the equilibrium condition of the differentials of the amount of substance of the material in solid state dn^s and the solution dn^{sl}

$$dn^s + dn^{sl} = 0. \tag{1.13}$$

This condition is met if there is no chemical potential gradient at the phase interface that would lead to the predominance of one of the flows I_{out} or I_{in}. Such equilibrium state is characterized by the zero value of the Helmholtz energy differential $dF = 0$. Thus, at the planar interface ($r \rightarrow \infty$) in equilibrium, the field of chemical potential is homogeneous, and the chemical potential of the substance in the solid $\mu^s(\infty)$ is equal to the chemical potential $\mu^{sl}(\infty)$ of the substance in solution

$$dF = \mu^{sl}(\infty)dn^{sl} + \mu^s(\infty)dn^s = 0 \rightarrow \mu^{sl}(\infty)dn^s = \mu^s(\infty)dn^s. \quad (1.14)$$

For a spherical nanoparticle of radius r and surface S_r, the differential of the Helmholtz energy dF is supplemented by the work of the interfacial tension σ^{sl} at the differential change of the surface $dS_r = 8\pi r dr$

$$\begin{aligned} dF &= \mu^{sl}(r)dn^{sl} + \sigma^{sl}dS_r + \mu^s(\infty)dn^s = 0 \\ \mu^{sl}(r)dn^s &= \mu^s(\infty)dn^s + \sigma^{sl}dS_r \end{aligned}. \quad (1.15)$$

The following applies to the differential increase of the amount of substance of solid phase dn^s

$$dn^s = \frac{dV_s}{V_m^s} = \frac{4\pi r^2 dr}{V_m^s} \rightarrow dr = \frac{V_m^s}{4\pi r^2}dn^s \quad (1.16)$$

and the differential Eq. (1.15) then assumes a simple integral form

$$\mu^{sl}(r) - \mu^s(\infty) = \frac{2\sigma^{sl}V_m^s}{r}. \quad (1.17)$$

Assuming that the activity is equal to the concentration $a = c$, the difference between the two chemical potentials can be written in the form

$$RT\ln c(r) - RT\ln c(\infty) = \frac{2\sigma^{sl}V_m^s}{r} \quad (1.18)$$

and by simple mathematical modification, we obtain the final formula for the saturated concentration of the substance above the surface of the nanoparticle with radius r

$$c(r) = c(\infty)e^{\frac{2\sigma^{sl}V_m^s}{RTr}}, \quad (1.19)$$

which is usually called the Ostwald–Freundlich equation [12]. According to this equation, as the radius r decreases, the solubility of the solid nanoparticle substance in the surrounding solution increases. The consequence of this fact is, for example, the existence of a critical size of the nucleation nucleus of a new phase, above which the particles are already stable, and they do not dissolve back into the solution.

This behavior is also the cause of a phenomenon called Ostwald ripening, which could be characterized at the best by biblical verse "For to everyone who has will be given, and he will have abundance, but from him who doesn't have, even that which he has will be taken away."$_{\text{Matthew 25.29}}$. To illustrate the Ostwald ripening process, consider a solution of concentration c (r_0) that is in equilibrium with a monodisperse system of particles of radius r_0.

$$c(r_0) = c(\infty)e^{\frac{2\sigma^{sl}V_m^s}{RTr_0}} \qquad (1.20)$$

Such a liquid dispersion of particles is very stable under the neglect of statistical fluctuations of radius r_0. If we keep the total weight of the dispersion fraction, but divide it into smaller particles with radius $r_1 < r_0$ and larger particles with radius $r_2 > r_0$, then both parts of the system are in imbalance. While smaller particles with radius $r_1 < r_0$ are surrounded by an unsaturated solution with concentration $c(r_0)$, smaller than their equilibrium concentration $c(r_1)$

$$c(r_1) = c(\infty)e^{\frac{2\sigma^{sl}V_m^s}{RTr_1}} \geq c(r_0), \qquad (1.21)$$

and a larger fraction with radius $r_2 > r_0$ is surrounded by a supersaturated solution with concentration $c(r_0)$, larger than their equilibrium concentration $c(r_2)$

$$c(r_2) = c(\infty)e^{\frac{2\sigma^{sl}V_m^s}{RTr_2}} \leq c(r_0). \qquad (1.22)$$

The solid material dissolves in the unsaturated solution and increases in size in the supersaturated solution. This leads to a gradual reduction of the small particles to complete dissolution and enlargement of the large particles by further crystallization from the supersaturated solution. In the final phase, only large particles with the complete absence of small ones remain in the dispersion.

1.1.2.3 Chemical Reactivity of Nanoparticles

As the particle size of one reactant, homogeneously dispersed in the liquid medium of the other reactant (gas or liquid), decreases, an increase in the reaction rate of this heterogeneous reaction is observed at their same volume concentration. In addition to the usual reaction parameters of macroreactions such as the volume concentration of both reactants, there are two basic factors that affect the reaction rate in the field of nanoparticles:

(1) the total reaction surface of both reactants per unit volume and (2) the reaction rate based on the reaction surface unit, and (1) when the radius r of spherical particles from a material of density ρ decreases, their specific surface increases according to the dependence

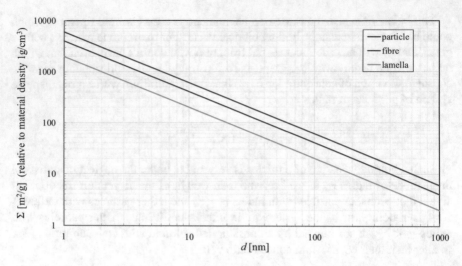

Fig. 1.4 Theoretical dependences of specific surfaces $\Sigma(d)$ of spherical particles according to (1.23), circular fiber of diameter d or lamellae of thickness d. The graphs confirm a significant increase $\Sigma(d)$ in the nanoscale region below 100 nm

$$\Sigma(r) = \frac{S(r)}{M(r)} = \frac{4\pi r^2}{\frac{4}{3}\pi r^3 \rho} \rightarrow \Sigma(r) = \frac{3}{r\rho}. \tag{1.23}$$

Figure 1.4 illustrates on the example of the unit density of the material not only the dependence of the specific surface $\Sigma(r)$ of a spherical particle but also analogous dependences for fibers and lamellae.

Surface area increases indirectly proportional to the size of the particles. Due to the choice of unit density, illustrative graphs on Fig. 1.4 can be also viewed in such sense that numerically at the same time they are surfaces related to a unit volume of 1 cm^{-3}, $\Sigma(d)$ (m^2cm^{-3}). For a simple combination reaction

$$mA + nB \rightarrow A_m B_n \tag{1.24}$$

on the surface of the nanoparticle of solid reactant A, surrounded by liquid reactant B, the reaction rate can be expressed by the kinetic equation for the reaction rate v

$$v = k[A]^m[B]^m. \tag{1.25}$$

Both parameters of molar concentrations $[A]$, $[B]$ have a macroscopic quantitative character and the stoichiometric parameters m, n in turn directly represent the internal characteristics of the $A_m B_n$ molecule, given by its quantum chemical behavior. Thus, the dimensional dependence of the reaction rate remains to be found in the macroscopic reaction rate constant $k = k(r)$. If we analyze the problem of reactivity in simple conditions of linear law (1.25) without extreme concentrations, then it can

Table 1.3 Dimensional dependence of the parameters values decisive for the rates of heterogeneous chemical reactions on the surface of particles [15]

r	$\Sigma(r)$ $(m^2 g^{-1})$	$G_s(r)$ (J)]
0.5 nm	$\approx 10^3$	$\approx 10^3$
0.5 μm	$\approx 10^1$	$\approx 10^0$
0.5 mm	$\approx 10^{-2}$	$\approx 10^{-3}$

be stated that doubling the reaction area $S \rightarrow 2S$ also doubles the reaction rate. This confirms the direct proportionality of the macroscopic reaction rate constant to the reaction area, relative to one mole $\Sigma_{mol}(r)$ of solid particulate material

$$v = \underbrace{k_1(r) \cdot \Sigma_{mol}(r)}_{k} \cdot [A]^m [B]^m. \qquad (1.26)$$

Then the parameter $k_1(r)$ in Eq. (1.26) represents the reaction rate constant of the reaction, relative to the unit area, which no longer contains a quantitative dependence on the size of the reaction area. Table. 1.3 shows the basic dependence of the specific surface area of particles Σ and the surface Gibbs energy G_s on the radius r of the particles.

The term reactivity should not be characterized by extensive parameters such as the reaction area, but should be understood at a deeper microscopic level of the molecules themselves and their immediate surroundings, such as the problem of the difference between homogeneous and heterogeneous nucleation of a new phase in precipitation reactions. For these reasons, the microscopic reaction rate constant $k_1(r)$, and its dependence on the change in the Gibbs energy of the reaction system ($\Delta G = G_{output} - G_{input}$), is decisive for the changes in reactivity depending on the particle size, and (2) next, let us analyze the reaction rate constant $k_1(r)$ in Eq. (1.26) related to the unit area of the reaction surface. It should be emphasized here that this is also a complex parameter, which depends on the morphology of the phase interface of both reactants. (*Truly fundamental characteristics of reactivity tend to be obtained at the purely molecular level by studying crossed molecular beams and are not related to the size of nanoparticles.*) To find the dependence of the reaction rate constant $k_1(r)$ on the size of the nanoparticle, it is advantageous to use Eyring equation

$$k_1(r) = \frac{RT}{N_A h} e^{-\frac{\Delta G(r)}{RT}}, \qquad (1.27)$$

which was derived by quantum mechanical methods based on the theory of the activated complex. Since the absolute temperature T is a fundamental thermodynamic variable and the universal gas constant R, the Avogadro constant N_A and the Planck constant h cannot contain a functional dependence, the magnitude dependence must be contained in the surface component of the Gibbs energy $G_s(r)$ related to the reaction phase interface. This can be expressed using the surface energy γ and the molar surface Σ_{mol} [16, 17] as follows:

$$G_s = \gamma \Sigma_{\text{mol}} \rightarrow G_s(r) = \gamma \frac{3}{r} V_m. \tag{1.28}$$

To express a specific dependence on the radius r in Eq. (1.27), let us start from the following situation. In the initial state, characterized by the Gibbs energy G_1, the reaction system is formed by a dispersion of nanoparticles of reactant A in a liquid solution of reactant B, and both components of the reaction are in the exact stoichiometric ratio. In the final state, characterized by the Gibbs energy G_2, the reaction system consists only of a solution of the reaction product $A_m B_n$. The Gibbs end-state energy consists only of the volume component $G_2 = G_{\text{bulk}}(A_m B_n)$ of the $A_m B_n$ reaction product solution without any phase interface. In contrast, the Gibbs energy of the initial state is composed of three components

$$G_1(r) = G_{\text{bulk}}(A) + G_{\text{bulk}}(B) + \underbrace{G_s(r)}_{\gamma \frac{3}{r} V_m}. \tag{1.29}$$

To change the Gibbs energy $\Delta G(r) = G_{\text{výstup}} - G_{\text{vstup}} = G_2 - G_1(r)$ in Eyring Eq. (1.27), with respect to the limit $\lim_{r \to \infty} G_s(r) = 0$, the dependence on the radius r of the nanoparticle can be separated from the reaction change of the Gibbs energy

$$\Delta G(r) = \Delta G(\infty) - \gamma \frac{3}{r} V_m. \tag{1.30}$$

Based on the validity of the relation (1.30) and Eyring Eq. (1.27), we obtain the final relation of the dependence of the reaction rate constant $k_1(r)$ on the radius

$$k_1(r) = k_1(\infty) e^{\frac{3\gamma V_m}{RTr}}. \tag{1.31}$$

From this it is evident that for large values of the radius of ordinary macroscopic particles $r \to \infty$, the reaction rate constant acquires a constant value $k_1(r)$ which no longer depends on the size; see Fig. 1.5.

To change $\Delta G < 0$, the reaction proceeds spontaneously, from the initial state with a higher value of Gibbs energy $G_1(\infty) > G_2$ and the reduction of the particle size, and this process significantly supports $G_1(r) > G_2$.

This situation is very effectively illustrated in Fig. 1.6, in which iron nanoparticles, manufactured by NANOIRON, s.r.o., ignite in a fall simply by interaction with air.

1.1.3 Electromagnetic Properties of Nanoparticles

1.1.3.1 Local Plasma Oscillation

When the light with wavelength λ interacts with nanoparticles, which have dimension smaller than λ, several new phenomena occur that do not occur with macroscopic

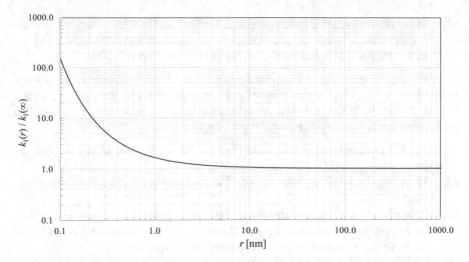

Fig. 1.5 Dependence of the ratio of coefficients $k_1(r)/k_1(\infty)$ in (1.31) calculated for the model value $3\gamma V_m/RT = 0.5$

Fig. 1.6 As an example of increased reactivity, these photographs of iron nanoparticles, which ignite only due to free fall, can be used [18]

objects. At the quasi-classical level, it is the effect of forced oscillations of the electron gas around relatively immobile ions of the crystal lattice of metals, known as local plasma oscillations. When the light of wavelength λ interacts with metal nanoparticles, whose dimensions are smaller than λ, it causes the following:

(1)   The electric intensity vector of the electromagnetic wave is always approximately constant throughout the volume of the particle. The particle thus behaves as if it was placed in a homogeneous electric field, which is harmonically dependent on time.

(2)   If the electron density in the metal conduction band is high enough to meet the requirement of the central limit theorem of statistical thermodynamics [4], we can consider it as a real "electron gas" where waves can be formed.

While the Sommerfeld quantum model must be used to describe some phenomena in the electron gas, in the description of electron oscillations in metal nanoparticles, we can suffice with the classical description within Drude model. This classical approximation is based on the following assumptions:

(1)   Electron momentum changes only during collisions with positively charged ions of the crystal lattice.
(2)   Electrons do not interact between collisions (independent electron approximation).
(3)   This model does not consider the crystal lattice potential. There is not field of other particles which would act on the electrons between collisions, and they move with a constant velocity (free electron approximation).
(4)   Collisions with positively charged ions are instantaneous and lead to a change in the momentum of electron.
(5)   Electron gas can reach thermodynamic equilibrium with the environment only through collisions with positively charged ions.
(6)   Electrons collide with frequency τ^{-1}, where τ is mean time between two collisions, at room temperature approximately 10^{-14} s (relaxation time approximation).

Within this approximation, let us now analyze the collective motion of an electron gas in a metallic environment which is under the influence of an external homogeneous electric field with intensity $\mathbf{E}$. This field acts on individual electrons with a force $\mathbf{F}_0 = -e\mathbf{E}$ and causes a redistribution of the collective electron gas density to a new equilibrium configuration in a way that the induction $\mathbf{D}$ inside the metal is zero

$$\varepsilon_0\mathbf{E} + \mathbf{P} = \mathbf{D} = 0 \tag{1.32}$$

and zero is therefore the total force acting on the electron $\mathbf{F} = 0$. In this configuration, the electron gas is maintained with newly established balance of the external force $\mathbf{F}_0$ and the oppositely oriented polarization force $\mathbf{F}_P = (e/\varepsilon_0)\mathbf{P}$

$$\mathbf{F}_0 + \mathbf{F}_P = \mathbf{F} = 0. \tag{1.33}$$

This polarization force $\mathbf{F}_P$ compensates the effect of external force by internal movement of electron gas of mean density n_e at a distance $x = \|\mathbf{x}\|$, which corresponds to the polarization $\mathbf{P} = -n_e e\mathbf{x}$, and its magnitude is expressed by the relation

$$\mathbf{F}_P = -\frac{n_e e^2}{\varepsilon_0}\mathbf{x}. \tag{1.34}$$

Static Eq. (1.33) then takes a specific form

$$F = -e\mathbf{E} - \frac{n_e e^2}{\varepsilon_0}\mathbf{x} = 0. \tag{1.35}$$

For further analysis of the electron gas, we perform a one-dimensional simplification $\mathbf{x} = x\mathbf{e}_1$ for the movement of electrons only along the x-axis, so Eq. (1.35) can then be written in a simple scalar form

$$F = -eE - \frac{n_e e^2}{\varepsilon_0}x = 0. \tag{1.36}$$

After the "immediate shutdown" of the external field E, it becomes a dynamic description of the undamped harmonic oscillator

$$m_e \frac{d^2 x}{dt^2} = F = -\frac{n_e e^2}{\varepsilon_0}x. \tag{1.37}$$

The solution of Eq. (1.37) is the function $x(t) = x_0 e^{-i\omega_p t}$, describing the undamped linear oscillations around the equilibrium point x_0 in the direction of this position vector. The value of ω_p comes from the structure of Eq. (1.37) and corresponds to the so-called plasma frequency of natural collective oscillations of the electron gas in the metallic material

$$\omega_p^2 = \frac{n_e e^2}{m_e \varepsilon_0}. \tag{1.38}$$

Now let us analyze the behavior of an electron gas under an external homogeneous harmonic electric field $E(t) = E_0 e^{-i\omega t}$ with amplitude E_0 and frequency ω. The field acts on the electron gas with a time-varying force

$$F(t) = -e E_0 e^{-i\omega t} \tag{1.39}$$

and it elicits a response in the form of forced oscillations. However, in the Drude model, electron collides with positive ions with a mean frequency of τ^{-1} during oscillations, which results in damping. Therefore, we must express the force F in Eq. (1.39) by using the first momentum approximation p based on the impulse theorem for the time $\Delta t = \tau$. The physical meaning of the time difference Δt arises from the analysis of the electron motion in the solid background of positive lattice ions. The change in the momentum Δp in one instantaneous collision cannot be realistically determined within the Drude model, so it is replaced by a change in momentum after two subsequent collisions, which can be related to the final mean time τ between collisions.

$$\left.\begin{array}{l} F\Delta t = \Delta p \\ \Delta p = p + \frac{dp}{dt}\Delta t \end{array}\right\} \rightarrow F = \frac{p}{\tau} + \frac{dp}{dt} \rightarrow F = \frac{m_e}{\tau}\frac{dx}{dt} + m_e\frac{d^2 x}{dt^2}. \tag{1.40}$$

The forced electrons oscillations are then described by the equation of motion with damping

$$m_e \frac{d^2 x}{dt^2} = -e E_0 e^{-i\omega t} - \frac{m_e}{\tau} \frac{dx}{dt}. \tag{1.41}$$

and the solution is a time-dependent oscillation deviation

$$x(t) = \frac{e}{m_e \left(\omega^2 + \frac{i}{\tau}\omega\right)} E_0 e^{-i\omega t}, \tag{1.42}$$

which cause time-varying electrical polarization

$$P(t) = -\frac{n_e e^2}{m_e \left(\omega^2 + \frac{i}{\tau}\omega\right)} E_0 e^{-i\omega t}. \tag{1.43}$$

Superposition of the external field and the polarization field then can express the time-varying induction

$$D(t) = \varepsilon_0 \underbrace{\left(1 - \frac{\omega_p^2}{\omega^2} \frac{1}{1 + \frac{i}{\omega\tau}}\right)}_{\varepsilon_r(\omega)} E_0 e^{-i\omega t}, \tag{1.44}$$

where the magnitude of the relative permittivity ε_r depends on the frequency ω of the external electric field

$$\varepsilon_r(\omega) = \underbrace{\left[1 - \frac{\omega_p^2}{\omega^2 + 1/\tau^2}\right]}_{\text{Re}(\varepsilon_r)} + i \underbrace{\left[\frac{\omega_p^2}{\omega\left(\omega^2 + 1/\tau^2\right)}\right]}_{\text{Im}(\varepsilon_r)}. \tag{1.45}$$

The system can be described by a quasi-static approximation if an electric field with a periodically varying intensity $E(t)$ is represented by an electric component of electromagnetic radiation with a wavelength much larger than the diameter of a metal nanoparticle $\lambda \gg 2R$. All electrons within the volume are subjected to the same external force (1.39) at the same time and, as a result they coherently oscillate around static equilibrium positions within the particle. This creates an electric dipole oscillator, which emits electromagnetic waves and affects the primary radiation field.

Subsequently, the modification of the electric field of the primary electromagnetic radiation, caused by the presence of a spherical metal nanoparticle, which is significantly smaller than its wavelength λ, is described.

Time-varying primary electric intensity $\mathbf{E}_0(t)$ acts on spherical nanoparticle (see Fig. 1.7), and its given potential $\Phi(t)$ is in a quasi-static approximation with fixed time t a solution of the Laplace equation with boundary conditions $\Delta\Phi(t) = 0$ [19]

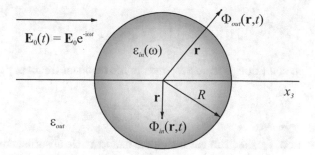

Fig. 1.7 Schematic of a spherical metal nanoparticle in electric field $\mathbf{E}_0(t)$ of an electromagnetic wave in the direction of the x_3 axis. While the relative permittivity of the environment has a constant value of ε_{out}, the relative permittivity of the particle ε_{in} shows a dependence (1.45) on the frequency ω of the incident wave. The inner $\Phi_{\text{in}}(\mathbf{r}, t)$ and outer $\Phi_{\text{out}}(\mathbf{r}, t)$ electric potential characterize the spatial distribution of the quasi-static field of electromagnetic radiation at time t, by the actual configuration of electrons oscillating inside the particle modified

$$\left.\frac{\partial}{\partial \theta} \Phi_{\text{in}}(\mathbf{r}, t)\right|_{r=R} = \left.\frac{\partial}{\partial \theta} \Phi_{\text{out}}(\mathbf{r}, t)\right|_{r=R}$$
$$\left.\varepsilon_{\text{in}} \frac{\partial}{\partial r} \Phi_{\text{in}}(\mathbf{r}, t)\right|_{r=R} = \left.\varepsilon_{\text{out}} \frac{\partial}{\partial r} \Phi_{\text{out}}(\mathbf{r}, t)\right|_{r=R} , \tag{1.46}$$

and is different for the region inside the particle $\Phi_{\text{in}}(t)$ and outside the particle $\Phi_{\text{out}}(t)$

$$\Phi_{in}(\mathbf{r}, t) = -\frac{3\varepsilon_{out}}{\varepsilon_{in}+2\varepsilon_{out}}(\mathbf{E}_0(t) \cdot \mathbf{r})$$
$$\Phi_{out}(\mathbf{r}, t) = -(\mathbf{E}_0(t) \cdot \mathbf{r}) + \frac{\varepsilon_{in}-\varepsilon_{out}}{\varepsilon_{in}+2\varepsilon_{out}}\left(\frac{R}{r}\right)^3(\mathbf{E}_0(t) \cdot \mathbf{r}) . \tag{1.47}$$

The electric intensity $\mathbf{E}_{\text{out}}(\mathbf{r}, t) = -\Delta\varepsilon_{out}(\mathbf{r}, t)$ (potential gradient outside the nanoparticle) then becomes superposition of the primary field $\mathbf{E}_0(t)$ and the external response of the induced field coming from the nanoparticle

$$\mathbf{E}_{\text{out}}(\mathbf{r}, t) = \mathbf{E}_0(t) + \frac{\varepsilon_{\text{in}} - \varepsilon_{\text{out}}}{\varepsilon_{\text{in}} + 2\varepsilon_{\text{out}}} \left(\frac{R}{r}\right)^3 \left(\frac{3}{r^2}\mathbf{r}(\mathbf{r} \cdot \mathbf{E}_0(t)) - \mathbf{E}_0(t)\right). \tag{1.48}$$

If the absolute value $|\varepsilon_{in}(\omega) + 2\varepsilon_{\text{out}}|$ goes to minimum, the second inductive term of the Formula (1.48) reaches a maximum, and the field is amplified by resonation. In the region of high frequencies (compared to the frequency of precipitation τ^{-1}), the imaginary part $\text{Im}(\varepsilon_{\text{in}}(\omega))$ is very small and the resonant maximum can be achieved by fulfilling the so-called Fröhlich condition for the formation of dipole localized surface plasmon on the nanoparticle

$$\text{Re}(\varepsilon_{\text{in}}(\omega)) = -2\varepsilon_{\text{out}}. \tag{1.49}$$

Maximizing the time τ of free movement of electron between two collisions with ions is the condition for the minimum absorption loss $\text{Im}(\varepsilon_{\text{in}}(\omega))$ around the resonant frequency

$$\text{Re}(\varepsilon_{\text{in}}(\omega)) = \lim_{\tau \to \infty} \left[1 - \frac{\omega_p^2}{\omega_0^2 + 1/\tau^2} \right] = 1 - \frac{\omega_p^2}{\omega_0^2}. \tag{1.50}$$

and the Fröhlich condition then corresponds to the resonant frequency ω_0

$$\omega_0 = \frac{\omega_p}{\sqrt{1 + 2\varepsilon_{out}}}. \tag{1.51}$$

In this situation, the electron gas inside the nanoparticle absorbs maximum electromagnetic energy by resonance, the surface charge density on opposite sides of the dipole electric oscillator changes periodically due to its oscillations, and the phenomenon is called localized surface plasmon resonance (LSPR). This dipole electric oscillator emits amplified electromagnetic radiation to the surroundings, which corresponds to the second term of Formula (1.48) for the outer field when the condition (1.50) is met. Formula (1.48) then becomes a dependence of the total electrical component of the radiation outside the nanoparticle $\mathbf{E}_{\text{out}}$ $(\mathbf{r}, t, \omega)$ on the frequency ω of the incident radiation

$$\mathbf{E}_{\text{out}}(\mathbf{r}, t, \omega) = \mathbf{E}_0(t) + \left(\frac{R}{r}\right)^3 \left(\frac{3}{r^2}\mathbf{r}(\mathbf{r} \cdot \mathbf{E}_0(t)) - \mathbf{E}_0(t)\right)$$
$$- \left[\frac{3\varepsilon_{\text{out}}}{1 + 2\varepsilon_{\text{out}} - \frac{\omega_p^2}{\omega^2}} \right] \left(\frac{R}{r}\right)^3 \left(\frac{3}{r^2}\mathbf{r}(\mathbf{r} \cdot \mathbf{E}_0(t)) - \mathbf{E}_0(t)\right) \tag{1.52}$$

While the first two terms (1.52) correspond to radiation, whose interaction with nanoparticles is frequency independent, the third term at the bottom contains in square brackets a frequency dependence characteristic LSPR. In practice, there are metallic materials with a high electron density such as gold (5.9×10^{22} cm^{-3}), silver (5.86×10^{22} cm^{-3}) or copper (8.47×10^{22} cm^{-3}) [20], for which the Fröhlich condition of resonant maximum in the visible region of the electromagnetic spectrum is fulfilled. This effect has been used in history, for example to color glass in temple stained glass or glass cups. A well-known example is the decorated cup of Thracian King Lycurgus from the fourth century A.D. (Fig. 1.8).

The Lycurgus Cup appears green in reflected light, but red in transmitted light. This is due to LSP resonance on nanoparticles of silver (66.2%), gold (31.2%) and copper (2.6%), which are homogenously dispersed in the glass matrix [22]. The different predominant chromaticity of the reflected and transmitted light called dichroism is caused in the case of transmitted red by the resonant absorption of the green component in the region of 520 nm. The green color, which dominates in the reflected light, is then the result of resonant scattering by silver nanoparticles in the region of the above frequencies.

Fig. 1.8 So-called Lycurgus Cup, probably made in Alexandria in the fourth century A.D., can serve as an illustration of the historical use of LSPR nanotechnology. On the left, you can see the appearance in reflected external light, on the right in transmitted light from a source inside the cup (Images used from http://www.britishmuseum.org [21])

1.1.3.2 Blue Shift of the Optical Absorption Spectrum

For very small nanoparticles approx. 2–50 nm (quantum dots), the effects of so-called quantum restriction are manifested. When reducing the dimensions of semiconductor nanoparticles, the so-called blue shift of the edges in the absorption spectra to shorter wavelengths occurs. This is a consequence of the quantum restriction which increases the width of the band gap [23–25], and it limits the electron energy spectrum and the mobility of the electron–hole exciton pair in the limited space of the nanoparticle material structure.

As the dimensions of nanoparticles decrease, the volume fraction of the boundary region of the crystal lattice gradually increases, in which Born–von Karman periodic boundary conditions of infinite periodicity are no longer well satisfied as shown in Fig. 1.9. This reduces the density of states at the boundaries of the energy bands and isolated energy levels appear; see Fig. 1.9. At the same time, quantum restriction also limits the energy of quasi-particles, such as excitons, that may be present inside the nanoparticle.

The width of the band gap corresponds to the energy required to excite an electron from the valence band into the conduction band and subsequently create two free charge carriers: an electron (e^-) in the conduction band and a hole (h^+) in the valence band. In this definition, both charges are considered at rest relative to the lattice and at the same time far apart so that their Coulomb interaction is negligible. However, if they approach each other, said Coulomb attraction can form a bound state, called the Wannier exciton, which can be roughly described by the Schrödinger equation with the Wannier Hamiltonian H_w, with a formal structure corresponding to the Hamiltonian for the hydrogen atom

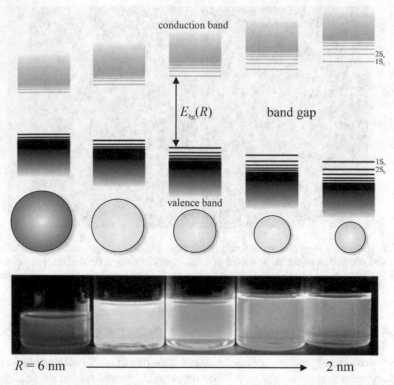

Fig. 1.9 Schematic illustration of the effect of quantum restriction on the fluorescence of CdSe (quantum dots). As the particle size decreases, not only does the band gap expand, but also the density of states at its boundaries decreases so significantly that isolated energy levels gradually separate, and their mutual distance gradually increases across clusters up to the limit of discrete spectrum of the isolated molecule. At the lower part, the fluorescence of a series of five CdSe nanodispersions excited by a UV lamp is shown. Image was taken and edited from [26]

$$\underbrace{\left[-\frac{\hbar^2}{2m_e^\circ}\Delta_e - \frac{\hbar^2}{2m_h^\circ}\Delta_h - \frac{e^2}{\varepsilon|\mathbf{r}_e - \mathbf{r}_h|} \right]}_{\hat{H}_w} \Psi(\mathbf{r}_e, \mathbf{r}_h) = E\Psi(\mathbf{r}_e, \mathbf{r}_h). \tag{1.53}$$

The characteristic value of the binding energy of an exciton in a semiconductor is in the order of hundreds of electron volts, and its spin can be equal to zero in the case of a singlet and equal to one in the case of a triplet (with three projection values: −1, 0, + 1). However, compared to the stability of a hydrogen atom, the lifetime of an exciton is in the order of nanoseconds. In contrast to the hydrogen atom, the relatively high permittivity $\varepsilon \gg 1$ in the Coulomb potential corresponds to the polarization effects in the respective semiconductor material, and the magnitudes of the effective masses of electrons m_e° and holes m_h° are much smaller than the masses of real electrons and protons. This combination of small masses of bound particles (e^-) and (h^+) and weak Coulomb attraction causes the extension of the exciton wave function over a relatively large region of the lattice structure.

In a simple quantum mechanical model of a hydrogen atom, there exists for the ground state ($n = 1$) a quantity called the Bohr radius [27, 28]

$$a_B = \frac{4\pi \varepsilon_0 \hbar^2}{e^2 m_e},$$ (1.54)

which represents the most probable distance of an electron from a central proton, and in a certain sense, it directly characterizes the dimension of the atom in the ground state.

With regard to the accepted analogy with the hydrogen atom and the influence of the crystal lattice of the surrounding material ($\varepsilon > \varepsilon_0$), the Bohr radius of the exciton a_{ex} can be expressed by a formally identical formula

$$a_{ex} = \frac{4\pi \varepsilon \hbar^2}{e^2 M_{ex}} = \frac{4\pi \varepsilon \hbar^2}{e^2} \left(\frac{1}{m_e^\circ} + \frac{1}{m_h^\circ} \right),$$ (1.55)

in which the mass of the electron m_e is replaced by the reduced mass of the exciton M_{ex}. It also characterizes the "exciton dimension." The lowest value of the "ground state" of the electron excitation energy corresponds to the width of the band gap, and the Bohr radius of the exciton has, for example, a diameter of about 6 nm for CdS. It is this dimensional characteristic in confrontation with the radius R of the nanoparticle that is important for the existence of the exciton inside the material.

In the quantum mechanical description of an exciton inside the crystal lattice of a material, the radius of the nanoparticle must be much larger than the lattice constant $R \gg a$ in order to preserve (unlike the cluster) the periodic structure. (56) Under these conditions, the exciton of the nanoparticle describes a wave function consisting of the product of the Bloch function $\psi_{bloch}(\mathbf{x})$ (exciton center of gravity) and the function $\varphi_{rest}(\mathbf{x})$, which describes the correction of exciton motion under the influence of quantum dimensional restriction $\Phi(\mathbf{x}) = \psi_{bloch}(\mathbf{x}) \cdot \varphi_{rest}(\mathbf{x})$. Schrödinger Eq. (1.56)

$$\left[\frac{\hbar^2}{2M_{ex}} \Delta_{r,\theta,\varphi} - V(r) \right] \varphi_{rest}(r, \theta, \varphi) = E \varphi_{rest}(r, \theta, \varphi)$$ (1.56)

describes the motion of center of gravity of the exciton in a spherical rectangular potential well ("spherical box" of radius R), defined by the potential

$$V(r, \theta, \varphi) = \begin{cases} V_0 < 0 \text{ for } r \leq R \\ V_0 = 0 \text{ for } r > R \end{cases}.$$ (1.57)

The custom function of Schrödinger's Eq. (1.56) can be written in a separate form as the product of radial Bessel functions $R(r)$ and spherical harmonics $Y_l^m(\theta, \varphi)$

$$\varphi_{rest}(r, \theta, \varphi) = R(r) \cdot Y_l^m(\theta, \varphi). \tag{1.58}$$

For its own lowest value of the energy of the ground state of the exciton "trapped" by a potential well (1.57) in the volume of the nanoparticle, it applies

$$E_{ex}(R) = \frac{\hbar^2 \pi^2}{2R^2 M_{ex}} = \frac{\hbar^2 \pi^2}{2R^2} \left(\frac{1}{m_e^\circ} + \frac{1}{m_h^\circ} \right). \tag{1.59}$$

If the formal radius of the exciton (1.55) is larger than the radius of the nanoparticle $a_{ex} > R$, it can be figuratively said that "the exciton does not fit into the material as a bound quasi-particle." Such a state of an electron and a hole corresponds to a so-called strong quantum restriction, and both the electron and the hole move in the crystal lattice without significant mutual coupling. The excitation energy required to produce such a state therefore includes the energy of the band gap $E_g^{bulk}(\infty)$, the dominant (kinetic) energy of the pair e^- and h^+, their effective Coulomb interaction and the last contribution correspond to their correlation energy proportional to the Rydberg energy E_{Ry}^{ex} exciton

$$E_g^{nano}(R) = E_g^{bulk}(\infty) + \frac{\hbar^2 \pi^2}{2R^2} \left(\frac{1}{m_e^\circ} + \frac{1}{m_h^\circ} \right) - 1.786 \frac{e^2}{8\pi \varepsilon R} - 0.248 E_{Ry}^{ex}. \tag{1.60}$$

The quantity $E_g^{nano}(R)$ is formally referred to as the band gap of a nanoparticle of radius R and Eq. (1.60), first published by Louis Brus [29, 30], bearing his name. Application of the formula for Rydberg exciton energy

$$E_{Ry}^{ex} = \frac{e^4}{32\pi^2 \varepsilon^2 \hbar^2 \left(\frac{1}{m_e^\circ} + \frac{1}{m_h^\circ} \right)} \tag{1.61}$$

and the Bohr radius of the exciton a_{ex} (1.55), Brus's Eq. (1.60) can be reformulated into the form of a general dependence of the relative change in energy $\delta E = (E_g^{nano}(R) - E_g^{bulk}(\infty))/E_{Ry}^{ex}$ on the ratio of the nanoparticle radius to Bohr's exciton radius $x = R/a_{ex}$

$$\delta E(x) = \left(\frac{\pi}{x} \right)^2 - \left(\frac{1,786}{x} \right) - 0.248. \tag{1.62}$$

Graphically, Brus's Eq. (1.62) is illustrated in Fig. 1.10 for three variants of the effectiveness of individual effects.

The effects of Coulomb interaction and correlation should be understood not as complementary. Their negative signs lead to a weakening of the restriction effect, which completely eliminates it in the range of values $3 < x < 6$. This occurs in cases of weak restriction, when the "exciton fits into the material of the nanoparticle" $R \gg a_{ex}$, and its kinetic energy begins to be comparable to the energy of the Coulomb interaction and the correlation energy.

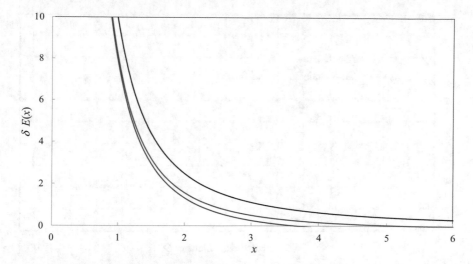

Fig. 1.10 Graphical characteristics of the effect of quantum excitation energy restriction in nanoparticles according to the general Brus Eq. (1.62). The black curve describes the dependence of $\delta E(x)$ for the case of strong restriction when the influence of the second two terms of the equation in relation to the first (kinetic energy) is negligible. The blue curve describes the dependence of $\delta E(x)$ for the case of weak restriction at lower kinetic energy, when the Coulomb interaction effect is already applied, and the red curve describes the dependence of $\delta E(x)$ taking into account the correlation effect [31]

The effect of quantum restriction, described by the Brus equation, has very interesting applications, and one of them is the determination of nanoparticle dimensions and quantum dots by measuring the absorption edge by analysis of UV–VIS absorption spectra [32, 33].

1.1.3.3 Superparamagnetism of Nanoparticles

As the particle size of the ferromagnetic material decreases to the region of the characteristic nanodimensions of their magnetic domains (<100 nm), a more energetically advantageous single-domain structure is formed at a certain level below the critical dimension D_c (see Fig. 1.11).

The border between single-domain and multidomain states is determined by the critical diameter

$$D_c \sim \frac{\gamma}{\mu_o M_s^2}. \tag{1.63}$$

which is directly proportional to the areal energy density γ of the domain wall where M_s is the spontaneous magnetization of the nanoparticle material [35].

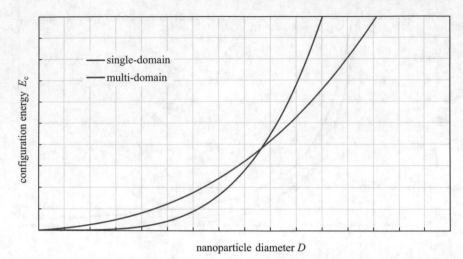

Fig. 1.11 Comparison of energy dependencies of single-domain and multidomain configurations. Below the intersection of the two curves (in the spherical approximation, it corresponds to the critical diameter D_c), the single-domain configuration is more energetically advantageous, while above the already multidomain configurations are created [34]

At the appropriate dimensional level, the magnetic behavior of a particle is well described by a "macrospin approximation." In this approximation, the magnetic moment of a nanoparticle has the character of a superposition of parallel magnetic moments of all atoms in its volume [35]. These single-domain ferromagnetic nanoparticles are magnetically anisotropic. Their preferential magnetization is realized in alternative antiparallel directions along the "easy axis." The two antiparallel directions of magnetization are separated by the energy barrier E_b. Table. 1.4 shows several examples of critical dimensions D_c for different materials.

Because the height of the energy barrier E_b is proportional to the volume V of the particle ($E_b = K_a V$), it becomes comparable to the mean energy of the thermal motion of atoms (K_a—coefficient of anisotropy). With its sufficient size, there can be an accidental collective change in the magnetization of the entire volume due to thermal fluctuations [15]. This happens randomly during short time intervals with

Table 1.4 Critical dimensions of D_c single-domain particles of magnetic materials

Material	D_c (nm)
SmCo$_5$	750
γ-Fe$_2$O$_3$—maghemite	166
Fe$_3$O$_4$—magnetite	128
Ni	55
Fe	15
Co—FCC	7

the mean size known as Neel's relaxation time τ_N in the range of 10^{-9} až 10^{-12} s [36, 37]. The magnitude is given by the Neel–Arrhenius equation with a time constant τ_0 of order 10^{-9}

$$\tau_N = \tau_0 e^{\frac{E_b}{k_B T}} \tag{1.64}$$

and its energy barrier height E_b plays the role of activation energy here. If the magnetic polarization measurement takes place without the presence of an external magnetic field for a time τ_m, much longer than the Neel relaxation time $\tau_N \ll \tau_m$, the polarization states compensate with each other in time, and the average magnetization shows zero value [38]. Otherwise, the much shorter measurement time $\tau_m \ll \tau_N$ results in the so-called blocking of the current polarization state, which at the time of measurement is not subsequently compensated by the opposite polarization. Under normal laboratory conditions, the characteristic measurement time (interaction) is $\tau_m \approx 100$ s and can be considered as a constant parameter. The boundary between the ferromagnetic and blocked state of the particle is then not determined by the mentioned measurement time, but by the so-called blocking temperature T_B, which is defined by the condition $\tau_N = \tau_m$

$$\ln \frac{(\tau_N = \tau_m)}{\tau_0} k_B T_B = E_b \rightarrow \ln \underbrace{\frac{100}{10^{-9}}}_{25} k_B T_B \approx E_b \rightarrow T_B \approx \frac{E_b}{25 k_B} \tag{1.65}$$

The blocking energy barrier is proportional to the volume of the particle $E_b = K_a V$; therefore, we can also derive from Eq. (1.65) the formula for the maximum limit volume V_{sp} (dimension D_{sp}) of the particle with blocking temperature T_b

$$V_{sp} = \frac{25 k_B T_B}{K_a} \rightarrow D_{sp} \approx \left(\frac{25 k_B T_B}{K_a} \right)^{\frac{1}{3}}. \tag{1.66}$$

The very small dimensions of D_p also correspond to the low-temperature limits of T_B below the level of normal laboratory temperatures. Under these conditions, the internal kinetic energy exceeds the blocking energy barrier and creates a state without remanent magnetization. In the external magnetic field, the activation energy increases from the original magnitude of E_b by the difference of the potential energies $\Delta E_{\uparrow\downarrow}$ of the two antiparallel polarizations. A set of these nanoparticles is magnetized without hysteresis as a paramagnetic material. This phenomenon is called super-paramagnetism and, compared to classical paramagnetic behavior, is characterized by significantly higher values of magnetization [39, 40], as shown graphically in Fig. 1.12.

The role that the magnetic moments of individual atoms play in classical para-magnets is taken over by coherently polarized nanoparticles in superparamagnets, and their magnetic moments reach high values of up to 10^4 Bohr magnetons ($\mu_B = 9.27 \cdot 10^{-24}$ J·T^{-1}).

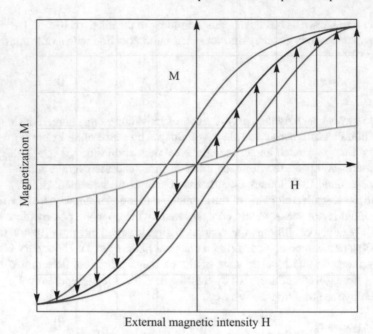

Fig. 1.12 Unlike macroscopic particles of ferromagnetic material—which are characterized by
a hysteresis loop the superparamagnetic magnetization curve—shows zero coercive field value.
However, it shows much higher values of saturated magnetization than standard paramagnetic
material—

Subsequently, we must analyze the effect of the external magnetic field on the
total activation energy in Eq. (1.64). Uniaxial anisotropy of particles (dominant axis
of multiple symmetry) has a significant effect on magnetization. If the main axis
of multiple symmetry is parallel with the axis from spherical coordinates and the
magnetization makes an angle θ with the axis, then the volume energy density ε
of the anisotropic magnetization outside the z direction is proportional to the $\sin \theta$
square in the first approximation [41]

$$\varepsilon \sim \sin^2\theta \rightarrow \frac{E_b}{V} = \varepsilon = K_a\sin^2\theta \rightarrow E_b = K_aV\sin^2\theta. \qquad (1.67)$$

External magnetic induction **B** acts on the magnetic moment **m** of the particle
and changes the original activation energy by a contribution

$$E_m = -\mathbf{m} \cdot \mathbf{B} = -MV\mu_0H\cos\theta, \qquad (1.68)$$

where M is the magnetization of the particle and H is the intensity of the external
field. The total value of the activation energy is then according to (1.67) and (1.68)
the function of orientation, characterized by the angle θ

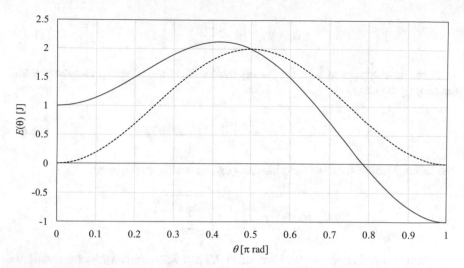

Fig. 1.13 Illustrative dependences of the polarization energy of the particle on the orientation θ for the case with—139 (69) and without—(67) the presence of an external field

$$E = E_b + E_m \rightarrow E(\theta) = K_a V \sin^2\theta - MV\mu_0 H\cos\theta. \tag{1.69}$$

For the polarization transition from the parallel orientation $\theta = 0$ to the opposite orientation $\theta = \pi$, antiparallel to the intensity of the external field, the system must exceed the energy maximum $E(\theta_{max})$ on Fig. 1.13.

While without the external magnetic field, the energy maximum corresponds to the orthogonal configuration $\theta = \pi/2$, and the presence of the external magnetic field brings asymmetry to the whole description; see Fig. 1.13. In the case of a parallel orientation $\theta = 0$, the energy barrier ΔE corresponds to a reverse polarization

$$\Delta E = E_{max} - E_{min} = E(\theta_{max}) - E(0) \tag{1.70}$$

While the energy level of a state with parallel orientation comes from a simple substitution $\theta = 0$ to (1.69)

$$E(0) = -MV\mu_0 H, \tag{1.71}$$

for the θ_{max} position and the corresponding maximum $E(\theta_{max})$, we must apply a local extreme condition

$$\frac{d}{d\theta}E(\theta_{max}) = 0 \rightarrow \begin{cases} \cos\theta_{max} = -\frac{M\mu_0 H}{2K_a} \\ E(\theta_{max}) = K_a V + K_a V \left(\frac{M\mu_0 H}{2K_a}\right)^2 \end{cases} \tag{1.72}$$

After the (1.71) and (1.72) substitution to (1.70), we can use the coercive intensity H_c to express the height of the energy barrier ΔE

$$\Delta E = K_a V \left(1 - \frac{M \mu_0 H_c}{2 K_a} \right)^2 . \tag{1.73}$$

After its comparison with the energy barrier E_b at the blocking temperature T_B according to (1.65)

$$K_a V \left(1 - \frac{M \mu_0 H_c}{2 K_a} \right)^2 = 25 k_B T_B \tag{1.74}$$

the corresponding coercive field intensity can be written in the form.

$$H_c = \frac{2 K_a}{M \mu_0} \left(1 - \sqrt{\frac{25 k_B T_B}{K_a V}} \right) \tag{1.75}$$

For macroparticles $V \to \infty$, Formula (1.75) provides an intensity of the coercive field H_{c0} for a bulk and can be reduced to

$$H_{c0} = \lim_{V \to \infty} H_c = \frac{2 K_a}{M \mu_0} \to H_c = H_{c0} \left(1 - \sqrt{\frac{25 k_B T_B}{K_a V}} \right) \tag{1.76}$$

Due to the relation for limit volume V_{sp} of a superparamagnetic particle (1.66), we can express the dependence of the coercive field H_c on the size D as follows:

$$H_c = H_{c0} \left(1 - \sqrt{\frac{V_{sp}}{V}} \right) \to H_c \approx H_{c0} \left(1 - \left(\frac{D_{sp}}{D} \right)^{\frac{3}{2}} \right), \text{for} D \in \left(D_{sp}, D_c \right) \tag{1.77}$$

The real dependence $H_c(D)$ reaches its maximum in the upper limit of the interval D_c and decreases asymptotically to a constant value of H_{c0} for macroscopic bodies (see Fig. 1.14). The function (1.77) ceases to be a good approximation in this area, because it reaches a maximum at the asymptotic limit $D \to \infty$ and not at the point of critical size D_c.

Near the critical size D_c, there are single-domain ferromagnetic particles with a significantly higher magnitude of the coercive field intensity than in macroscopic multidomain material. We can create exceptionally strong permanent magnets with a high remanent field by fixing them into a dense solid particle dispersion.

Another very interesting application of the superparamagnetism phenomenon is a colloidal dispersion of those nanoparticles in a liquid (water or non-polar solvents), which are called "magnetic liquids" or "ferrofluids."

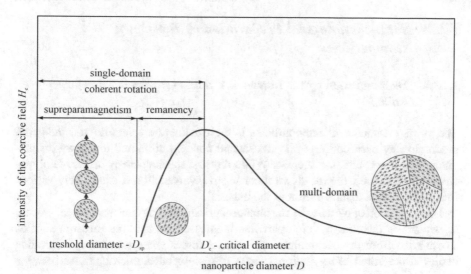

Fig. 1.14 Dependence of the particle coercive field intensity H_c on its diameter D and on its division into zones with different magnetization behavior [42]. An example of a very narrow hysteresis loop for Fe_2O_3 nanoparticles with a mean size of 100 nm is shown in the upper right corner [43]

1.2 Origin of Nanoparticles

1.2.1 Natural Formation of Nanoparticles

In the primordial stages of the evolution of the material form of matter, the first nuclei of future massive bodies were formed by the condensation of atoms scattered into the cosmic vacuum by supernova explosions. This "matter condensation" initially created most of the atomic clusters, which gradually grew in a bottom-up process into larger objects, and some may still be part of cosmic dust.

Probably, the most significant alternative process for the natural formation of nanoparticles is the erosive disintegration of more massive particles entrained in the flow of gas or liquid on the surface of planets during their multiple collisions.

1.2.2 Anthropogenic Formation of Nanoparticles as Waste

A very serious and growing problem today is the parasitic genesis of nanoparticles in various forms of a by-product of human activity. Industrial fumes, combustion products, friction braking products and many other sources have been an unintentional source of nanoparticles for over a hundred years [44]. At present, this group is a significant environmental burden, which is slowly and often very laxly limited.

1.2.3 Nanotechnological Preparation of Top-Down Nanoparticles

1.2.3.1 Disintegration by the Impact of Grinding Bodies in Planetary Mills

The youngest source of nanoparticles is their deliberate scientific and industrial production by both bottom-up synthesis and condensation, and top-down disintegration, or a combination thereof. While the first mechanism is based mainly on chemical processes, the top-down mechanism involves almost exclusively various forms of physical disintegration of the material.

Disintegration of particles by the impact of grinding bodies is a mechanical process in which the material of microparticles is stressed by extreme pressure shocks. These arise from the accumulation of inertial forces when macroscopic grinding bodies strike small cross-sectional areas of disintegrated microparticles between them (Fig. 1.15).

The chamber is inserted into the peripheral position of the impeller of the mill and performs its own rotational movement against the direction of rotation during each rotation. Such a "planetary" movement at high speed (up to 1100 rpm) leads to very intensive mixing of the entire filling. Grinding bodies are usually in the form of balls made of very hard and abrasion-resistant, high-density material. Intensive

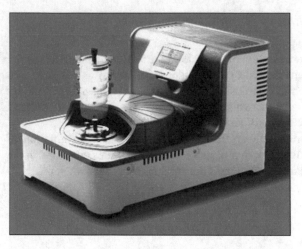

Fig. 1.15 High-performance planetary mill PULVERISETTE 7 premium line from company FRITSCH GmbH—milling and sizing [45] with a grinding limit of 0.1 μm. Interchangeable grinding chambers with a volume of up to 80 ml (right) are filled with grinding balls with a diameter of 0.5–20 mm made of tungsten carbide (or agate, corundum, silicon nitride, zirconium oxide or hardened steel) as required

Fig. 1.16 Laboratory pearl nanomill NETZSCH microseries [46] with an input fraction size of about 0.02 mm and an output fraction reaching levels below 100 nm. In the right part, the arrangement of the grinding chamber and the rotary agitator can be seen, the speed of which reaches a maximum of 4200 rpm

disintegration into the size of nanoparticles is in most cases realized by the above-mentioned multiple impacts between hard balls in the grinding chamber during its opposite rotational movement.

1.2.3.2 Disintegration by Mutual Impacts of Grinding Pearls and Impact on Moving Parts of Fast Rotor in Pearl Mills

In disintegration agitators, the particles of primary material are dispersed in a liquid dispersion along with a large number of miniature balls ("grinding pearls") of hard abrasion-resistant material. These are mixed at high speed with a hard rotor, creating a large number of intense collisions. In this disintegration technology, the material density of the hard rotor is not important, because the impulses during collisions with its surface are transmitted to the particles by its stiffness directly from the drive unit (Fig. 1.16).

In the process of disintegration by mutual grinding of particles between grinding pearls, the high concentration of their representation in the volume of the grinding chamber plays a key role, which guarantees extreme frequency of mutual collisions and high efficiency of disintegration.

1.2.3.3 Disintegration by Mutual Impacts During Entrainment into the Fast Gas Stream and Impacts on the Hard Walls of the Grinding Chamber

In jet mills, particles of disintegrated material are entrained at high speed by a carrier gas and circulate in a disk disintegration chamber in a cyclone mode (Fig. 1.17).

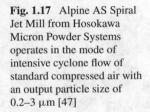

Fig. 1.17 Alpine AS Spiral
Jet Mill from Hosokawa
Micron Powder Systems
operates in the mode of
intensive cyclone flow of
standard compressed air with
an output particle size of
0.2–3 μm [47]

During their circulation, there are very intense mutual collisions and at the same time collisions with the hard walls of the chamber. By cyclone separation, the smaller particles migrate ever closer to the axis of rotation until they are finally led out by an axial outlet pipe for the removal of compressed air and further separated by filtration.

1.2.3.4 Disintegration by Ultrasonic Cavitation in Liquid Suspension

In cavitation disintegration, particles of disintegrated material are dispersed in a liquid suspension that circulates in a very strong ultrasonic field. An example of the technical application of an ultrasonic field in a liquid in the cavitation disintegration of small solid particles is the ultrasonic cavitation disintegrator (see Fig. 1.18).

By heterogeneous nucleation, sedentary cavitation bubbles are preferably formed on the surface of the particles (see Fig. 1.19), which by subsequent implosion disintegrate the particles by impact pressures up to GPa units.

In direct contact with the liquid dispersion of particles, there is only a titanium sonotrode and mostly a stainless steel flow chamber. Due to the fact that the nucleation of cavitation bubbles occurs mainly on the highly curved surface of dispersed particles, and the contamination of the ground material by cavitation damage of the sonotrode material and the chamber walls is significantly lower than, for example, jet mills.

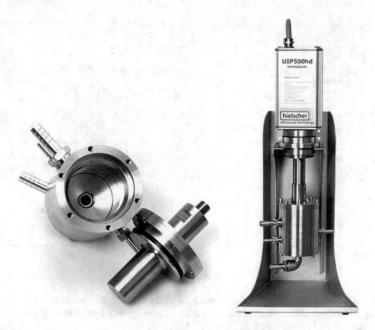

Fig. 1.18 Example of an ultrasonic flow disintegrator UIP500hd from company Hielscher Ultrasonics GmbH [48]. In the upper part of the assembly on the right side, there is an electromechanical excitation module, which is terminated at the bottom by a cylindrical titanium sonotrode. This can also be seen on the left side of the picture in the lid of the disassembled disintegration chamber

Fig. 1.19 Scheme of the dominant disintegration mechanism of water jet mill (WJM [49]). There is a pulse effect of high impact pressures (water hammer effect) during the asymmetric collapse of the cavitation microbubble, seated on the surface of the particle

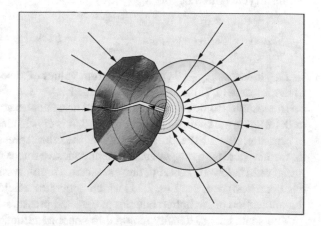

While in mixed steam-gas cavitation, the bubble collapse is inhibited by adiabatic compression of the released gas (air) to form a plasma cluster "hotspot," in the collapse of a purely steam bubble, this phenomenon of significant deceleration does not occur, and most of the energy is concentrated to the point of final impact of the water interface on the particle surface. At normal absolute pressure of liquid (0.1 MPa) and a temperature of 20 °C, the impact pressure of the hydraulic shock

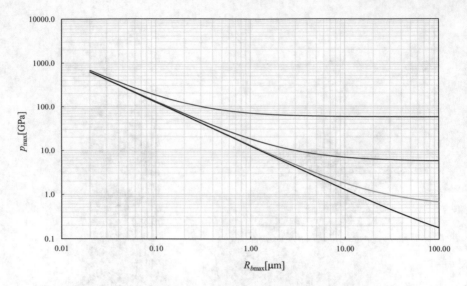

Fig. 1.20 Dependences of the impact pressure p_{max} on the initial maximum dimension R_{bmax} of the collapsing cavitation bubble. From above gradually for the absolute pressures of the liquid medium $p_o = -1$ MPa, -10^{-1} MPa, -10^{-2} MPa, -10^{-3} MPa

reaches a minimum value of 5.7 GPa. Said impact pressure for the aquatic environment depends on the absolute pressure of the liquid p_o and the initial radius of the collapsing cavitation bubble R_{bmax} [50]

$$p_{max}(p_o, R_{bmax}) = A p_o + \frac{B}{R_{bmax}}, \quad (A = 56895, B = 12414 \text{Pa} \cdot \text{m}). \qquad (1.78)$$

Dependences p_{max} (R_{bmax}) for different values of absolute pressure po according to the formula (1.78) are illustrated at Fig. 1.20.

In most ultrasonic disintegrators, cavitation bubbles collapse at a normal pressure of 0.1 MPa in the chamber. From Fig. 1.20, it is crystal clear that at a tenfold increase in pressure, e.g., 1 MPa, the magnitude of the impact pressure is already 57 GPa, and its further increase by the contribution of cohesive forces occurs only in the region of initial radii below 100 nm. For these reasons, a high-pressure ultrasonic disintegrator has been designed, see Fig. 1.21, which generates a field of sufficient intensity for the nucleation of cavitation bubbles at elevated pressures.

The sonic energy density required to generate ultimate tensile stresses in a fluid increases significantly with its external pressure p_o. Under these conditions, the nucleation of cavitation bubbles is more difficult, but their collapse has the potential to disintegrate even very hard materials.

Fig. 1.21 Example of a
high-pressure ultrasonic
disintegrator, which was
made to order by company
ULTRASONIC.CZ s.r.o. The
ultrasonic field is excited in a
2000 ml pressure chamber
by 16 Langevin's radiators
with a total output of 2 kW
(an octagonal housing with
fans is used to cool them).
The maximum pressure in
the chamber can reach up to
25 MPa and significantly
increases the impact pressure
of the hydraulic shock during
the implosion of cavitation
bubbles on the surface of
disintegrated particles

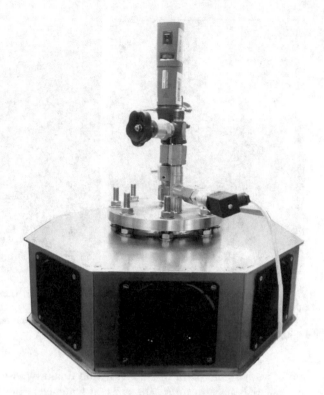

1.2.3.5 Disintegration by Cavitation of a High-Energy Liquid Jet in a Liquid Suspension

In the new type of nanodisintegrator water jet mill (WJM) [49], such a technological arrangement is used during grinding, in which the high-energy water jet emerges from the diamond primary nozzle at a speed of over 660 ms^{-1} into the environment of the aqueous suspension of ground particles in the high-cavitation mode. In the WJM workspace, the particles are entrained by the liquid suspension and cyclically pass through a system of disintegration zones generated by a cavitating liquid jet in a series of narrow "abrasive tubes" made of tungsten carbide.

The disintegration process in WJM uses extreme dynamics and high power density in the areas of interaction of the liquid jet with the suspension of ground material particles. At a working pressure of the pump of approx. 420 MPa and a flow speed of 660 ms^{-1}, the power density at the cross section reaches 164 kWmm^{-2}. With a water jet diameter of 230 μm, this corresponds to a value of 8 kW over the entire cross section of the jet.

The source of cavitation in the given technical arrangement is the already mentioned high velocity gradient and tensile stresses when mixing a fast liquid stream with a slow dispersion of ground particles. Asymmetric collapse of the cavitation bubble on the particle surface, see Fig. 1.22, leads to the formation of a very strong

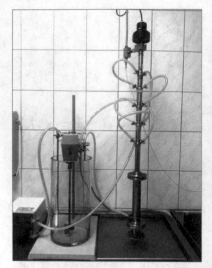

Fig. 1.22 On the left, a photo of the water jet mill disintegrator (patent—VŠB-Technical University of Ostrava [49]) with a limit in the area below 50 nm and on the right a high-pressure water multiplier up to 420 MPa by company PTV spol. s.r.o. [51] to supply the water jet mill disintegrator with high-pressure liquid up to 420 MPa

shock wave at the point of impact of the liquid, whose pressure maxima exceed the material strength limit. These listed parameters predetermine the principle of cavitation disintegration of WJM for application even on very hard materials with a reduced degree of contamination by structural elements of the disintegrator [52].

1.2.4 Nanotechnology Preparation of Nanoparticles by the Bottom-Up Method

1.2.4.1   Clustering in Vacuum and Shell Structure of Clusters

The method of condensation of atomic or molecular clusters in a vacuum or in a dilute gas and their subsequent growth to the size of nanoparticles is most often used for the preparation of metal nanopowders [53, 54]. The theory of gas phase condensation for the production of metal nanopowders was first described as early as 1930. Heating the starting solid material (usually in the form of an element heated by an electric current or arc, by laser ablation or melt emission in a Knudsen chamber) in a low vacuum containing inert or reactive gas causes its vaporization and cooling during expansion between gaseous molecules. In the case of an inert environment, condensation occurs into atomic clusters and their subsequent growth. In the case of a reactive gas, such as oxygen, there is first a reactive transformation of the evaporated atoms (oxidation in the case of oxygen) and subsequent nucleation and growth, as in the first case of simple condensation. In this process, there is also a significant proportion of heterogeneous reaction on the phase surface of already formed nanoparticles of the compound. The described cooling of metal vapors by gas molecules makes it possible to prepare globular liquid nanoparticles up to the size of nanometer units. These grow both by the described flow of atoms or molecules from the environment to their surface, and by mutual coalescence, when due to their fluidity, the globular shape is still preserved. After solidification by gradual cooling with gas molecules, their size no longer changes significantly.

1.2.4.2 Nucleation and Growth of a New Phase in Gas and Liquid

Two basic methods are used for the preparation of nanoparticles by precipitation of an insoluble new phase in liquids. One is to control the kinetics of precipitation reactions either by rapid depletion of one of the reactants [32, 33] or by blocking the reaction mechanism after reaching a specific value of the selected parameter (e.g., the formation of a micellate envelope around the nanoparticles) [55]. A separate method is to limit the amount of one reactant in a very small reaction volume as drops in microemulsion or aerosol methods of the liquid–liquid type [56, 57]. It is their size and the concentration of reactant inside that can control the future size of the nanoparticles, as a product of the precipitation reaction in a limited volume.

1.2.4.3 Nucleation and von Weimarn Rule

In classical precipitation reactions with controlled kinetics, the quality of the final product is governed by von Weimarn's rule [58–60]:

> The insoluble matter formed in the precipitation reaction is precipitated in the form of a colloidal dispersion (lyosol) if the starting materials have been mixed in very small or very high concentrations. In the middle concentration range, a coarse-grained precipitate is formed.

For the final particle size of the precipitate, its solubility in a liquid solvent, which in most cases is water, is decisive.

Under the low degree of supersaturation conditions shown in Fig. 1.23 approximately above the region of 0-A, the reaction product is relatively little concentrated in the form of a new insoluble phase and the description is problematic here.

In the region above section A-B, the resulting supersaturation is only sufficient to form crystalline nuclei, but their growth is limited by the low ambient concentration of the reaction product. A highly dilute fine dispersion of new phase particles is formed.

With a medium degree of supersaturation in the region above the section B-C, good conditions arise for both nucleation and growth of crystalline nuclei, and therefore a coarse-grained dispersion of the reaction product is formed here.

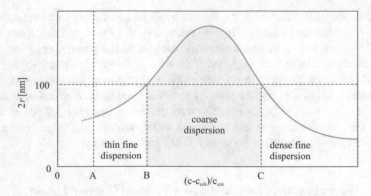

Fig. 1.23 Schematic course of the dependence of the mean radius *r* of the dispersion particles on the relative supersaturation of the precipitation reaction product (taken and adapted from [60], where *c* denotes the actual concentration and c_{sat} the saturation concentration of the precipitation reaction product)

At high supersaturation values in the region behind point C, the rate of nucleation significantly dominates over the rate of further growth of the formed nuclei. This situation occurs due to very extensive precipitation and a sharp decrease in the global concentration of product in the solvent. This significantly reduces the concentration gradient around the surface of the particles and thus the diffusion flux, which is necessary for further growth. The resulting dispersion system has up to a gel-like character. And precisely in connection with this quality of the reaction product, one of the most widespread methods of nanoparticle preparation is applied—the sol–gel method [61, 62].

1.2.4.4 Nucleation and Growth in Supercritical Fluid

Supercritical fluids in high-pressure reactors are also successfully used to prepare nanoparticles [63, 64]. The selected material, dissolved in a supercritical fluid (most often supercritical carbon dioxide, or supercritical water), is converted at an elevated temperature and pressure to a state above the critical point of the phase diagram, where there is no liquid–gas interface. Subsequent rapid expansion results in a very rapid constitution of the phase interfaces between small areas and the environment due to mass fluctuations in many places. These then become the nuclei of nanoparticles.

References

1. International Organization for Standardization (ISO), *Technical Specification (ISO/TS) 27687:2008; Nanotechnologies—Terminology and Definitions for Nano-objects—Nanoparticle, Nanofibre and Nanoplate*; First published 2008–08–15." [Online]. Available: https://www.iso.org/store.html.

2. O. Kvítek, P. Slepička, O. Lyutakov, V. Švorčík, Nanostruktury zlata: příprava, vlastnosti a vybrané aplikace. Chem. List. **110**, 922–930 (2016)
3. T.P. Martin, Shells of atoms. Phys. Rep. **273**(4), 199–241 (1996). https://doi.org/10.1016/0370-1573(95)00083-6
4. C. Vaz, *Thermodynamics and Statistical Mechanics*. University of Cincinnati.
5. The first nanogram balance sets - Sartorius History (1971–2000). https://www.sartorius.com/en/company/about-sartorius-ag/history/1971-2000.
6. M.J.D. Clift, K. Clift, G. Hunt, P. Gehr, B. Rothen-Rutishauser, NanoImpactNet Nomenclature. Eur. Netw. Heal. Environ. Impact Nanomater. v3 (2010)
7. F.A. Lindemann, The calculation of molecular vibration frequencies. Phys. Zeitschrift **11**, 609–612 (1910)
8. F.G. Shi, Size dependent thermal vibrations and melting in nanocrystals. J. Mater. Res. **9**(5), 1307–1314 (1994). https://doi.org/10.1557/JMR.1994.1307
9. K.J. Klabunde, (ed.), *Nanoscale Materials in Chemistry* (Wiley, 2001)
10. S.L. Lai, J.Y. Guo, V. Petrova, G. Ramanath, L.H. Allen, Size-dependent melting properties of small tin particles: nanocalorimetric measurements. Phys. Rev. Lett. **77**(1), 99–102 (1996). https://doi.org/10.1103/PhysRevLett.77.99
11. K.-J. Hanszen, Theoretische Untersuchungen uber den Schmelzpunkt kleiner Kugelchen. Zeitschrift fur Phys. **157**(5), 523–553 (1960). https://doi.org/10.1007/BF01340711
12. J. Leitner, M. Kamrádek, Termodynamický popis nanosystémů. Chem. List. **107**, 606–613 (2013)
13. J.W. Gibbs, J. Tyndall, *On the Equilibrium of Heterogeneous Substances : First [-Second] Part.* (New Haven: Published by the Academy, 1874)
14. G. Kaptay, The Gibbs equation versus the Kelvin and the Gibbs-Thomson equations to describe nucleation and equilibrium of nano-materials. J. Nanosci. Nanotechnol. **12**(3), 2625–2633 (2012). https://doi.org/10.1166/jnn.2012.5774
15. W.F. Brown, Thermal fluctuations of a single-domain particle. Phys. Rev. **130**(5), 1677–1686 (1963). https://doi.org/10.1103/PhysRev.130.1677
16. J. Leitner, D. Sedmidubský, Teaching nano-thermodynamics: gibbs energy of single-component nanoparticles. World J. Chem. Educ. **5**(6), 206–209 (2018). https://doi.org/10.12691/wjce-5-6-4
17. B.J. Block, S.K. Das, M. Oettel, P. Virnau, K. Binder, Curvature dependence of surface free energy of liquid drops and bubbles: A simulation study. J. Chem. Phys. **133**(15), 154702 (2010). https://doi.org/10.1063/1.3493464
18. "Nanopowder, consisting of Fe(0) nanoparticles 'NANOFER 25P.'" https://nanoiron.cz/en/products/zero-valent-iron-nanoparticles/nanofer-25p.
19. S.A. Maier, *Plasmonics: Fundamentals and Applications* (Springer, New York, NY US, 2007)
20. A. Bansal, S.S. Verma, Searching for alternative plasmonic materials for specific applications. Indian J. Mater. Sci. **2014**, 1–10 (14AD). https://doi.org/10.1155/2014/897125
21. "Lycurgus cup—exhibit of the British Museum." https://www.britishmuseum.org/research/collection_online/collection_object_details.aspx?objectId=61219&partId=1.
22. D.J. Barber, I.C. Freestone, An investigation of the origin of the colour of the Lycurgus cup by analytical transmission electron microscopy. Archaeometry **32**(1), 33–45 (1990). https://doi.org/10.1111/j.1475-4754.1990.tb01079.x
23. F.T. Rabouw and C. de Mello Donega, Excited-state dynamics in colloidal semiconductor nanocrystals. Top. Curr. Chem. **374**(5) 58 (2016), https://doi.org/10.1007/s41061-016-0060-0
24. C. Vatankhah, M. Saki, S. Jafargholinejad, Theoretical and experimental investigation of quantum confinement effect on the blue shift in semiconductor quantum dots. Orient. J. Chem. **31**(2), 907–912 (2015). https://doi.org/10.13005/ojc/310234
25. Y. Kayanuma, Quantum-size effects of interacting electrons and holes in semiconductor microcrystals with spherical shape. Phys. Rev. B **38**(14), 9797–9805 (1988). https://doi.org/10.1103/PhysRevB.38.9797
26. C. de M. Donegá, Synthesis and properties of colloidal heteronanocrystals. Chem. Soc. Rev. **40**(3), 1512–1546 (2011), https://doi.org/10.1039/C0CS00055H

27. R.P. Feynman, R.B. Leighton, M. Sands, *The Feynman Lectures on Physics, Vol. III: The New Millennium Edition: Quantum Mechanics* (Basic Books, 2011)
28. P.A.M. Dirac, *The Principles of Quantum Mechanics*, 3rd ed. (Oxford at the Clarendon Press, 1948)
29. L.E. Brus, Electron–electron and electron-hole interactions in small semiconductor crystallites: the size dependence of the lowest excited electronic state. J. Chem. Phys. **80**(9), 4403–4409 (1984). https://doi.org/10.1063/1.447218
30. L. Brus, Electronic wave functions in semiconductor clusters: experiment and theory. J. Phys. Chem. **90**(12), 2555–2560 (1986). https://doi.org/10.1021/j100403a003
31. R. Koole, E. Groeneveld, D. Vanmaekelbergh, A. Meijerink, C. de Mello Donegá, Size effects on semiconductor nanoparticles," in *Nanoparticles* (Springer, Berlin, Heidelberg, 2014), pp. 13–51
32. P. Praus, O. Kozák, K. Kočí, A. Panáček, R. Dvorský, CdS nanoparticles deposited on montmorillonite: preparation, characterization and application for photoreduction of carbon dioxide. J. Colloid Interface Sci. **360**(2), 574–579 (2011). https://doi.org/10.1016/j.jcis.2011.05.004
33. P. Praus, R. Dvorský, P. Horínková, M. Pospíšil, P. Kovář, Precipitation, stabilization and molecular modeling of ZnS nanoparticles in the presence of cetyltrimethylammonium bromide. J. Colloid Interface Sci. **377**(1), 58–63 (2012). https://doi.org/10.1016/j.jcis.2012.03.073
34. N.A. Spaldin, Analogies and differences between ferroelectricsand ferromagnets, in *Physics of Ferroelectrics: A Modern Perspective* (Springer Berlin Heidelberg, 2007), pp. 175–218
35. C. Caizer, *Nanoparticle Size Effect on Some Magnetic Properties, in Handbook of Nanoparticles* (Springer International Publishing, Cham, 2015), pp. 1–38
36. J. Dieckhoff, D. Eberbeck, M. Schilling, F. Ludwig, Magnetic-field dependence of Brownian and Néel relaxation times. J. Appl. Phys. **119**(4), 043903 (2016). https://doi.org/10.1063/1.4940724
37. R.J. Deissler, M.A. Martens, Y. Wu, R. Brown, Brownian and Néel relaxation times in magnetic particle dynamics, in *2013 International Workshop on Magnetic Particle Imaging (IWMPI)* (2013), pp. 1–1, https://doi.org/10.1109/IWMPI.2013.6528375
38. R.J. Deissler, Y. Wu, M.A. Martens, Dependence of Brownian and Néel relaxation times on magnetic field strength. Med. Phys. **41**(1), 012301 (2014). https://doi.org/10.1118/1.4837216
39. S.P. Gubin, (ed.), *Magnetic Nanoparticles* (Wiley, 2009)
40. R. Kodama, Magnetic nanoparticles. J. Magn. Magn. Mater. **200**(1–3), 359–372 (1999). https://doi.org/10.1016/S0304-8853(99)00347-9
41. C.L. Dennis et al., The defining length scales of mesomagnetism: a review. J. Phys. Condens. Matter **14**(49), R1175–R1262 (2002). https://doi.org/10.1088/0953-8984/14/49/201
42. G.M. Lekha, S. George, Colloidal magnetic metal oxide nanocrystals and their applications, in *Colloidal Metal Oxide Nanoparticles* (Elsevier, 2020), pp. 289–335
43. J. Lunacek et al., Efficiency of HIGH gradient magnetic separation applied to micrometric magnetic particles. *Sep. Sci. Technol.* 150701140517004 (2015), https://doi.org/10.1080/01496395.2015.1061006
44. J. Kukutschová, *Wear Particles from Automotive Brake Materials: Generation, Characterization, and Environmental Impact* (VŠB—Technical University of Ostrava, Ostrava, 2017)
45. FRITSCH Planetary Micro Mill PULVERISETTE 7—premium line. https://www.fritsch-international.com/sample-preparation/milling/planetary-mills/details/product/pulverisette-7-premium-line/downloads
46. NETZSCH—Laboratory bead Mill MiniSeries. https://www.netzsch-grinding.com/en/products-solutions/wet-grinding/miniseries-laboratory-mills
47. Hosokawa—Alpine AS Spiral Jet Mill. https://www.hmicronpowder.com/applications/size-reduction/alpine-as-spiral-jet-mill
48. Hielscher Ultrasonics GmbH—Ultrasonic mill UIP500hd. http://www.hielscher.com/ultrasonics/i500_p.htm.
49. R. Dvorský, Způsob dezintegrace pevných mikročástic do rozměrů nanočástic kavitujícím kapalinovým paprskem a zařízení k provádění tohoto způsobu, CZ 305704 (2016)

50. R. Dvorsky, J. Lunacek, A. Sliva, Dynamics analysis of cavitation disintegration of micropar-
    ticles during nanopowder preparation in a new Water Jet Mill (WJM) device. Adv. Powder
    Technol. **22**(5), 639–643 (2011). https://doi.org/10.1016/j.apt.2010.09.008
51. PTV, spol. s r.o. http://www.ptv.cz.
52. R. Dvorsky, J. Trojková, *Cavitation Disintegration of Powder Microparticles, in Handbook of
    Mechanical Nanostructuring* (Wiley-VCH Verlag GmbH & Co. KGaA, Weinheim, Germany,
    2015), pp. 533–549
53. K. Wegner, B. Walker, S. Tsantilis, S.E. Pratsinis, Design of metal nanoparticle synthesis by
    vapor flow condensation. Chem. Eng. Sci. **57**(10), 1753–1762 (2002). https://doi.org/10.1016/
    S0009-2509(02)00064-7
54. L. Xing, *Iron Nanoparticles by Inert Gas Condensation: Structure and Magnetic Characteri-
    zation* (University of Groningen, 2018)
55. R. Dvorsky, P. Praus, J. Trojkova, Model of synthesis of ZnS nanoparticles stabilised by
    cetaltrimethylamonium bromide. Chalcogenide Lett. **10**(10), 385–391 (2013)
56. R. Dvorsky et al., Synthesis of composite photocatalytic nanoparticles $ZnO \cdot mSiO_2$ using new
    aerosol method. Hut. List. **LXIX**, 68–72 (2016)
57. J. Bednář, L. Svoboda, P. Mančík, R. Dvorský, Synthesis of ZnS nanoparticles of required size
    by precipitation in aerosol microdroplets. Mater. Sci. Technol. **35**(7), 775–781 (2019). https://
    doi.org/10.1080/02670836.2019.1590514
58. P.P. von Weimarn, *Kolloid-Zeit. 3* (1908)
59. P.P. von Weimarn, *Kolloid-Zeit. 4* (1909)
60. P.P. von Weimarn, The precipitation laws. Chem. Rev. **2**(2), 217–242 (1925). https://doi.org/
    10.1021/cr60006a002
61. L.L. Hench, J.K. West, The sol-gel process. Chem. Rev. **90**(1), 33–72 (1990). https://doi.org/
    10.1021/cr00099a003
62. C. Aydın, H.M. El-Nasser, F. Yakuphanoglu, I.S. Yahia, M. Aksoy, Nanopowder synthesis of
    aluminum doped cadmium oxide via sol–gel calcination processing. J. Alloys Compd. **509**(3),
    854–858 (2011). https://doi.org/10.1016/j.jallcom.2010.09.111
63. N. Taniguchi, On the basic concept of nanotechnology, in *Proceedings of International
    Conference on Production Engingering Tokyo, Part II* (1974), pp. 18–23
64. K. Byrappa, S. Ohara, T. Adschiri, Nanoparticles synthesis using supercritical fluid tech-
    nology – towards biomedical applications. Adv. Drug Deliv. Rev. **60**(3), 299–327 (2008).
    https://doi.org/10.1016/j.addr.2007.09.001

Chapter 2
Interaction Among Nanoparticles

2.1 Weak van der Waals Interactions Among Molecules

In contrast to the much stronger bonding interactions between atoms and molecules (ionic, covalent, metallic), their weaker interactions are most significantly represented by the action of van der Waals forces. These arise by the interaction of different electric charges distributions within neutral atoms or molecules. Van der Waals interactions are manifested by the action of basic cohesive forces in condensed substances as well as adhesive forces that fix molecules on the surface of bodies during physical adsorption [1, 2]. In the dimensional region of nanoparticles and microparticles, a universal interparticle attraction arises between particles due to the integral action of van der Waals forces [2]. In the case of larger macroscopic objects, it manifests itself in nature as, for example, the adhesion of the limbs of geckos and many small insects.

The generally multipolar nature of the electric charge density distribution in neutral atoms and molecules is also time varying due to quasi-classical and quantum fluctuations. The electric potential of the i-th molecule $\varphi_i(\mathbf{r})$ is a scalar function of the spatial distribution of charge in the volume of the molecule and can be written in the general form of a multipolar expansion from the radial Coulombic potential of the total charge q_i with the longest range through the dipole moment field $\mathbf{p}_i$ and the quadrupole moment described by the second-order tensor $\mathbf{Q}_i$, to higher-order multipole moments.

$$\varphi_i(r) = \frac{1}{4\pi\varepsilon}\left(\frac{q_i}{r} + \frac{p_i \cdot r}{r^3} + \frac{1}{2}\frac{r \cdot Q_i \cdot r}{r^5} + \dots\right). \tag{2.1}$$

For overall neutral molecules, the first development term disappears, and the dipolar interaction term greatly exceeds the higher-order multipolar contributions

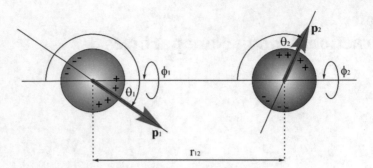

Fig. 2.1 Configuration of two different dipoles in spherical coordinates

$$\varphi_i(r) = \frac{1}{4\pi\varepsilon}\frac{p_i \cdot r}{r^3}.$$ (2.2)

In spherical coordinates according to Fig. 2.1, the potential energy U is the action of the potential $\varphi_1(\mathbf{r})$ of the dipole $\mathbf{p}_1$ on the electric dipole $\mathbf{p}_2$ and written as a function of the instantaneous values of the scalar coordinates $r_{12}(t) \rightarrow r_{12}$, $\theta_{1,2}(t) \rightarrow \theta_{1,2}$, $\phi_{1,2}(t) \rightarrow \phi_{1,2}$ and polarization $p_{1,2}(t) \rightarrow p_{1,2}$

$$U(p_1, p_2, r_{12}, \theta_1, \theta_2, \phi_1, \phi_2) = \frac{1}{4\pi\varepsilon}\frac{p_1 p_2}{r_{12}{}^3}(2\cos\theta_1\cos\theta_2 - \sin\theta_1\sin\theta_2\cos(\phi_2 - \phi_1)).$$ (2.3)

In different instantaneous redistributions of charges and currents, the electromagnetic interaction between molecules results in instantaneous forces that also fluctuate in time. The time mean values of these fluctuating forces then form the force field of the van der Waals interaction. The interaction between two electric dipoles can be divided into three basic categories:

1. $p_1 = $ const. and $p_2 = $ const.

Keesom interaction of two stable dipoles—two freely rotating stable dipoles preferentially orient their opposite charges toward each other during their random thermal motion due to mutual interaction, and thus mutual attraction prevails between them. Despite this fact, however, the other configurations, when moving thermally, cause a more pronounced decrease in potential energy with distance, which after centering over all orientations corresponds to a decrease with the sixth power of the mutual distance r_{12} [3]

$$U(p_1, p_2, r_{12}) = -\frac{C_{12}^{\mathrm{Kessom}}}{r_{12}{}^6} \quad \left[C_{12}^{\mathrm{Kessom}} = \frac{1}{3k_B T}\left(\frac{p_1 p_2}{4\pi\varepsilon}\right)^2 \right].$$ (2.4)

2. $p_1 = $ const. and $p_2 = p_2(t)$

Debye interaction of a permanent dipole with an induced dipole—when the charges of the permanent dipole of a polar molecule approach a non-polar molecule without a permanent dipole moment, the non-polar molecule undergoes a redistribution of charges due to the electric field of the permanent dipole, and an induced electric dipole is formed. Similar to the Keesom interaction, its orientation corresponds to the emergence of attractive forces relative to the original dipole of the first polar molecule. In addition, if both dipoles are freely rotatable, the forces decrease significantly with distance. The potential energy for the interaction between the permanent dipole and the induced dipole (with polarizability α_2) after centering over all orientations corresponds again to a decrease with the sixth power of the mutual distance r_{12} [4, 5]

$$U(p_1, \alpha_2, r_{12}) = -\frac{C_{12}^{\text{Debye}}}{r_{12}^6} \quad \left[C_{12}^{\text{Debye}} = \alpha_2 \left(\frac{p_1}{4\pi\varepsilon} \right)^2 \right]. \tag{2.5}$$

3. $p_1 = p_1(t)$ and $p_2 = p_2(t)$

London dispersion interaction of two fluctuating dipoles—in contrast to the Keesom and Debye interaction, a quantum mechanical approach is already needed to physically describe the interaction of two non-polar molecules. It is quantum mechanical perturbation theory that allows the description of the phenomenon we call dispersion forces, which corresponds to attractive interaction. As an illustration of the mechanism of formation of London dispersion, forces can serve the idea of an atom composed of a practically point atomic nucleus, around which electrons oscillate at their own energy levels with high frequencies (up to the vacuum UVC frequency of about 10 PHz). Thus, at each "sharply defined" instant, the molecules have a polar character, but their polarization changes with extremely high frequencies. The highest possible frequency of these oscillations ν_{max} for a given molecule is limited by the binding strength, which corresponds to its ionization energy $I = h\nu_{\text{max}}$. The respective electric fields of both oscillating dipoles represent external contributions to the electron Hamiltonians of both molecules and lead to a situation where the resulting instantaneous polarizations are more significantly correlated with the configurations of the corresponding attractions. And, as in both previous cases, the statistical centering across all orientations again corresponds to a decrease with the sixth power of the relative distance r_{12} [6]

$$U(\alpha_1, \alpha_2, I_1, I_2, r_{12}) = -\frac{C_{12}^{\text{London}}}{r_{12}^6} \left[C_{12}^{\text{London}} = \frac{3}{2} \frac{\alpha_1 \alpha_2}{(4\pi\varepsilon)^2} \left(\frac{I_1 I_2}{I_1 + I_2} \right) \right]. \tag{2.6}$$

The expression of the C_{London} parameter in (2.6) is a simplification due to the use of the ionization energies I_1 and I_2. These correspond only to the maximum oscillation frequencies ν_{max} which are the most dominant. For an accurate description, however, the entire excitation spectra of both molecules must be taken into account. In this context, it should be noted that the close connection between the nature of the interactions of the set of atoms in the condensed phase and oscillations of electrons at the respective levels provides the possibility of characterizing the interactions by measuring the absorption spectra of the material in the relevant region.

The simultaneous contribution of all three of the above effects gives rise to a general van der Waals interaction with the $C_{12}/r_{12}{}^6$ dependence which is responsible for the attractive part of the Lennard–Jones potential [2].

2.2 Adhesion Interactions Between Nanoparticles as a Result of a Collective van der Waals Interaction—Hamaker Microscopic Summation Method

Although the $C_{12}/r_{12}{}^6$ potential energy function shows a significant decrease with distance, at molecular scales and in the nanoscale region, the van der Waals interaction can be considered as a "long-range force." This fact naturally offered a generalization for the van der Waals interaction between macroscopic bodies at the relevant microscales, first published by Hamaker [7]. In the classical physics approximation, the additive nature of the electromagnetic interaction was used when the energy contributions U_{ij}^{vdW} of the interaction between any pair of molecules species 1 and 2 are independent of the influence of the surrounding molecules. The total U_{12} interaction energy between two particles 1 and 2 is then given by the simple summation of the contributions of all $N_1 \cdot N_2$ pairs of molecules between the total particle volumes v_1 and v_2

$$\left. \begin{array}{c} U_{12} = \sum_{i=1}^{N_1(v_1)} \sum_{j=1}^{N_2(v_2)} U_{ij}^{vdW} \\ U_{ij}^{vdW} = -\frac{C_{ij}}{r_{ij}{}^6} \end{array} \right\} \rightarrow U_{12} = -\sum_{i=1}^{N_1} \sum_{j=1}^{N_2} \frac{C_{ij}}{r_{ij}{}^6}. \qquad (2.7)$$

For the practical calculation of the total energy of the van der Waals interaction between two spherical particles 1 and 2 with radii R_1 and R_2, the individual molecules of formula (2.7) will be represented by small ensembles of molecules which, at constant bulk densities n_1 and n_2, are contained in differential volumes dv_1 and dv_2

$$\left. \begin{array}{c} \left. \begin{array}{c} \text{material 1 and 2} \\ \text{of particles 1 and 2} \end{array} \right\} \rightarrow C_{ij} = C_{12} \\ \left. \begin{array}{c} \text{volumetric particle} \\ \text{number density} \end{array} \right\} \rightarrow \frac{dN_i}{dv} = n_i \end{array} \right\} \rightarrow U_{12} = -\int_{v_1} \int_{v_2} \frac{C_{12}}{r_{12}{}^6} (n_1 dv_1)(n_2 dv_2).$$

$$(2.8)$$

The integral (2.8) represents the sum of the energies of all pairwise interactions between the molecules of particle 1 and particle 2. Its concrete form can be expressed using distances r_{12} and two spherical coordinate systems with origin O_1 and O_2 as shown in Fig. 2.2.

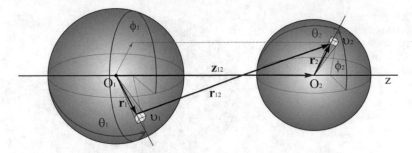

Fig. 2.2 Coordinate scheme for the integration of the total energy U_{12} of two spherical particles 1 and 2

To express the distance vector $\mathbf{r}_{12}$ between two interacting volume elements, it is convenient to use the vector identity

$$\mathbf{r}_1 + \mathbf{r}_{12} = \mathbf{r}_2 + \mathbf{z}_{12} \rightarrow \mathbf{r}_{12} = \mathbf{r}_2 - \mathbf{r}_1 + \mathbf{z}_{12}. \tag{2.9}$$

Based on the primary notation (2.9) using two Cartesian systems with origin O_1 and O_2

$$r_{12}{}^2 = (x_2 - x_1)^2 + (y_2 - y_1)^2 + (z_2 - z_1 + z_{12})^2. \tag{2.10}$$

the size of r_{12} can be formulated in the corresponding spherical coordinates with origins O_1 and O_2

$$r_{12}{}^2 = \left(\begin{array}{c} (r_1\sin\theta_1\cos\phi_1 - r_2\sin\theta_2\cos\phi_2)^2 \\ +(r_1\sin\theta_1\sin\phi_1 - r_2\sin\theta_2\sin\phi_2)^2 \\ +(r_1\cos\theta_1 - r_2\cos\theta_2 + z_{12})^2 \end{array} \right) \tag{2.11}$$

and the integral (2.8) then takes the form

$$U_{12} = C_{12}n_1n_2 \int_0^{R_1} dr_1 \int_0^{\pi} d\theta_1 \int_0^{2\pi} d\phi_1 \int_0^{R_2} dr_2$$

$$\int_0^{\pi} d\theta \int_0^{2\pi} \frac{r_1^2\sin\theta_1 r_2^2\sin\theta_{2_2}}{\left(\begin{array}{c} (r_1\sin\theta_1\cos\phi_1 - r_2\sin\theta_2\cos\phi_2)^2 \\ +(r_1\sin\theta_1\sin\phi_1 - r_2\sin\theta_2\sin\phi_2)^2 \\ +(r_1\cos\theta_1 + z_{12} - r_2\cos\theta_2)^2 \end{array} \right)^3} d\phi_2. \tag{2.12}$$

The integration (2.12) is extremely problematic from a mathematical point of view, and for these reasons, Hamaker's inventiveness in solving the problem very elegantly should be appreciated.

The first step is to determine the total interaction energy between all molecules of particle 1 and one outer molecule M at distance R from the center O_1; see Fig. 2.3.

Fig. 2.3 Scheme of
integration of interaction
energy contributions along
spherical differential layers
dr with constant distance
r from the molecule M

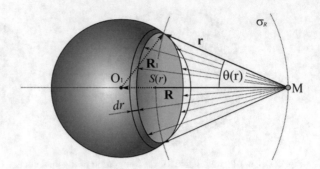

The aim is to obtain the radial dependence of the energy $U_1(R)$ of one molecule
M (material of particle 2) on the distance R from the center O_1 of particle 1. This is
the cumulative result of the van der Waals field of all radial differential layers of area
$S(r)$ over the entire volume of a particle of radius R_1. The area $S(r)$ of a spherical
cap with center M is expressed by the integration

$$S(r) = \int_0^{2\pi} d\phi \int_0^{\theta(r)} r^2 \sin\theta d\theta. \tag{2.13}$$

Since all molecules of the respective differential layer are at the same distance
r with respect to molecule M, the total energy of the van der Waals field can be
expressed by integration over the whole diameter of the particle 1

$$U_1(R) = -\int_{v_1} \frac{C_{12}}{r^6} (n_1 dv_1). \tag{2.14}$$

Expressing the volume differential dv_1 in terms of the area of the spherical cap
$S(r)$ (2.13)

$$\left. \begin{array}{l} dv_1 = S(r)dr \\ \mathbf{R_1} = \mathbf{R} - \mathbf{r} \rightarrow R_1^2 = R^2 + r^2 - 2Rr\cos\theta \\ r^2 = R_1^2 + 2Rr\cos\theta - R^2 \end{array} \right\} \rightarrow dv_1 = \pi \frac{r}{R} \left(R_1^2 - (R-r)^2 \right) dr \tag{2.15}$$

The integral (2.14) takes a particular form

$$U_1(R) = -\pi \frac{C_{12}n_1}{R} \int_{R-R_1}^{R+R_1} \frac{1}{r^5} \left(R_1^2 - (R-r)^2 \right) dr \rightarrow U_1(R)$$
$$= -\frac{4}{3}\pi R_1^3 C_{12}n_1 \frac{1}{\left(R^2 - R_1^2 \right)^3}. \tag{2.16}$$

In the following second step, we integrate the energy contributions of all molecules
of particle 2 in the $U_1(R)$ field. At this stage, particle 1 will be represented by the

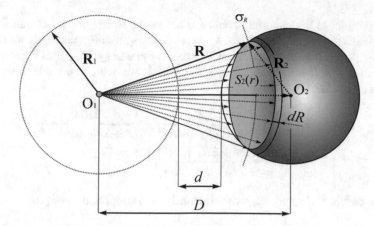

Fig. 2.4 Scheme of integration of interaction energy contributions along spherical differential layers dR with constant distance R from the center O_1 of particle 1

point source of the $U_1(R)$ field, and the integration of all energy contributions will be performed in reverse order analogous to the previous calculation; see Fig. 2.4.

The energy contribution of the van der Waals interaction of radial field $U_1(R)$ on all molecules of particle 2 on the sphere σ_R at distance R is equal to the $U_1(R)$ multiplied by the number of all molecules $(n_2 \cdot dv_2)$ in the differential layer dR

$$dU_{12}(R, D) = U_1(R) \cdot (n_2 \cdot dv_2(R)). \tag{2.17}$$

The total magnitude of the interaction energy $U_{12}(D)$ between the two particles with centers at distance D is obtained by integration over the entire volume of particle 2

$$
\left.
\begin{aligned}
U_{12}(D) &= \int_{D-R_2}^{D+R_2} U_1(R) \cdot (n_2 dv_2) \\
dv_2 &= \pi \frac{R}{D}\big(R_2^2 - (D-R)^2\big)dR \\
U_1(R) &= -\frac{4}{3}\pi R_1^3 C_{12} n_1 \frac{1}{(R^2 - R_1^2)^3}
\end{aligned}
\right\} \rightarrow U_{12}(D)
$$

$$
= -\frac{4\pi^2 R_1^3 C_{12} n_1 n_2}{3D} \int_{D-R_2}^{D+R_2} \frac{R\big(R_2^2 - (D-R)^2\big)}{\big(R^2 - R_1^2\big)^3}dR. \tag{2.18}
$$

After calculating the integral (2.18), we get

$$
U_{12}(D) = -\frac{1}{6}\pi^2 C_{12} n_1 n_2 \left[\frac{2R_1 R_2}{D^2 - (R_1 + R_2)^2} + \frac{2R_1 R_2}{D^2 - (R_1 - R_2)^2} \right.
$$

$$
\left. + \ln\left(\frac{D^2 - (R_1 + R_2)^2}{D^2 - (R_1 - R_2)^2} \right) \right]. \tag{2.19}
$$

The product of the material parameters before the square bracket is traditionally referred to as the Hamaker constant $A_{12} = \pi^2 C_{12} n_1 n_2$, usually specified for different materials [8, 9]. It is also customary to express the energy $U_{12}(d)$ as a function of the distance $d = D - R_1 - R_2$ between the particle surfaces, so the final interaction formula then takes the form

$$U_{12}(d) = -\frac{A_{12}}{6}\left[\frac{2R_1 R_2}{d(d + 2R_1 + 2R_2)} + \frac{2R_1 R_2}{(d + 2R_1)(d + 2R_2)}\right.$$
$$\left. +\ln\left(\frac{d(2R_1 + 2R_2 + d)}{(d + 2R_1)(d + 2R_2)}\right)\right].$$

(2.20)

In cases where $d \ll R_2, R_2$, we sometimes use a simplified form [10]

$$U_{12}(d) = -\frac{A_{12}}{6}\frac{1}{d\left(\frac{1}{R_1} + \frac{1}{R_2}\right)},$$

(2.21)

however, the agreement with the original formula (2.20) is already below 1.6% at a distance of two particle radii, so its application is very limited for most calculations. But if we fix the minimum contact distance, it provides a basis for calculating the static stability of the bound particles.

References

1. V.A. Parsegian, *Van Der Waals Forces: A Handbook for Biologists Engineers, and Physicists* (Cambridge University Press, Chemists, 2005)
2. J. Israelachvili, *Intermolecular and Surface Forces* (Elsevier, 2011)
3. W.H. Keesom, Die van der waalschen kohaesionkraft. Phys. Zeitschrift **22**(264), 12 (1921)
4. P. Debaye, *Phys. Z.* **21**, 178 (1920)
5. P. Debaye, *Phys. Z*, **22**, 302 (1921)
6. F. London, Zur Theorie und Systematik der Molekularkrafte. Zeitschrift fur Phys. **63**(3–4), 245–279 (1930). https://doi.org/10.1007/BF01421741
7. H.C. Hamaker, The London—van der Waals attraction between spherical particles. Physica **4**(10), 1058–1072 (1937). https://doi.org/10.1016/S0031-8914(37)80203-7
8. H.-J. Butt, M. Kappl, *Surface and Interfacial Forces* (Wiley-VCH Verlag GmbH & Co. KGaA, Weinheim, Germany, 2010)
9. L. Bergström, Hamaker constants of inorganic materials. Adv. Colloid Interface Sci. **70**, 125–169 (1997). https://doi.org/10.1016/S0001-8686(97)00003-1
10. R.J. Hunter, L.R. White, D.Y.C. Chan, *Foundations of Colloid Science*, no. sv. 1 (Clarendon Press, 1987)

Chapter 3
Self-Organization of Nanoparticles

3.1 Self-Organization of Nanoparticles

At present, the process of chemical self-organization of atoms and molecules into higher units with varying degrees of collective arrangement is generally very well known. Depending on the history of their solidification, solid natural materials show different degrees of arrangement, from the "short-range order" in amorphous glasses to the "long-range order" in large single crystals. Chemical elementary building elements have two very unique properties.

1. Atoms or molecules of a given element or compound are completely identical in terms of interactions.
2. The characteristic binding energy of the interaction between atoms or molecules exceeds the kinetic energy during collisions at normal temperatures, and they do not separate.

The consequence of these properties is an extremely diverse set of different self-organized units from snowflakes to large living organisms such as forests.

If we wanted to generalize the effect of self-organization for higher elements, such as nanoparticles, we find that the first-mentioned condition would significantly reduce the spectrum of self-organized structures. Practical experience shows that the identity condition is very insufficiently fulfilled for most nanoparticle sets. Most sets of nanoparticles are not strictly monodisperse, and their diverse shape is also a very strong limitation. Thus, only monodisperse sets of nanocrystalline particles with the same shape are shortlisted. Gold nanocrystals can serve as an example Fig. 3.1.

In addition to the above-mentioned second property, which is necessary for self-organization, the weight of the nanoparticles is also a very important factor. In many cases, the weight is too high for the attractive component of the mutual interactions to cause sufficient shifts to the approach necessary to form a stable bond. For the above reasons, the formation of a larger self-organized structure in a liquid dispersion is very problematic, where the Brownian motion is an important destructive factor already

© The Author(s), under exclusive license to Springer Nature Switzerland AG 2022
R. Dvorsky et al., *Nanoparticles' Preparation, Properties, Interactions and Self-Organization*, SpringerBriefs in Applied Sciences and Technology,
https://doi.org/10.1007/978-3-030-89144-2_3

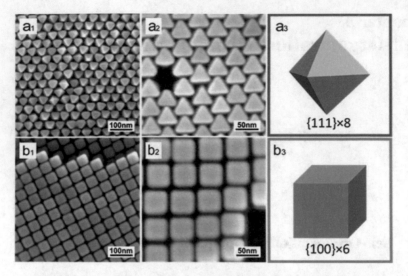

Fig. 3.1 SEM micrograph of regular self-organization of uniform gold nanocrystals. The left column contains a schematic representation of the corresponding polyhedral shapes—**a** Au-octahedral and **b** Au-cubic (Reprinted with permission from J. Am. Chem. Soc. 2017, 139, 21, 7310–7317; Publication Date: May 11, 2017; https://doi.org/10.1021/jacs.7b01735. Copyright 2017 American Chemical Society) [1]

at normal temperatures. Graphene and fullerene belong to a very modest group of nanoparticles that can be subjected to effective self-organization in accordance with the above conditions and will be discussed below.

A very special case of self-organization of nanometric elements is often a highly sophisticated biological organization into higher functional structures. This issue is a completely separate area and goes beyond the scope of this publication.

3.1.1 Graphene ↔ Graphite

Lamellar graphene nanoparticles can be included in the group of nanoparticles that meet the above-mentioned first identity condition. The term nanoparticle in this case defines smaller lamellae, which under certain conditions can be retrospectively oriented in parallel and bind by van der Waals to a multilayer parallel structure corresponding to graphite; see Fig. 3.2.

If graphene nanoparticles of sufficient density are dispersed in a liquid, they may gradually self-organize into multilayer graphene to graphite-like forms upon contact. Here, however, it should be noted that the opposite process of exfoliating the aggregated form of bulk graphite to structures with fewer layers is much more likely. An alternative is to join lamellas into larger lamellas over their edges. However, this possibility is much less likely due to unequal edge shapes.

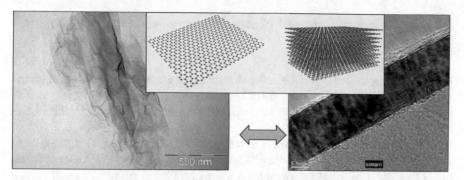

Fig. 3.2 TEM micrograph of a deformed graphene lamella (left) and TEM micrograph of a graphite multilayer section (right) (Reprinted from Chemical Physics Letters, Vol. 430, Issue 1–3, Prakash R. Somani, Savita P. Somani, Masayoshi Umeno; Planer nanographenes from camphor by CVD, Pages 56–59, Copyright 2006, with permission from Elsevier [2])

3.1.2 Fullerene ↔ Fullerite

At present, one of the most famous allotropic forms of carbon is a system of sixty atoms called fullerene. These carbons are covalently bound, and they create very solid globular form with a diameter of 0.71 nm, see Fig. 3.3 (left). Fullerene is a macromolecular structure, which is often already ranked among the nanoparticles. While carbon atoms in fullerenes are bound by strong covalent bonds, the individual fullerene particles react with each other by a weak van der Waals interaction. In these nanoparticles, this causes self-organization analogous to crystallization. Thus, regular aggregates–crystals of fullerite are formed; see Fig. 3.3 (right).

The solid mineral form of fullerite crystallizes in a planar face-centered cubic lattice (FCC) with a lattice constant of 1.416 nm. The fullerenes are relatively weakly bound in the system due to the dominance of the van der Waals interaction, which is the reason for its low hardness [3].

Fig. 3.3 Scheme of one fullerene nanoparticle C_{60} (left) and face-centered cubic cells of fullerite (right)

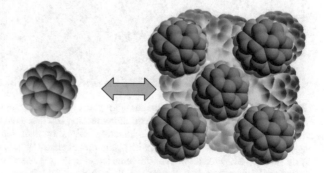

3.2 Self-Organization of Nanoparticles at the Sublimation Interface

In investigating the collective behavior of a set of large numbers of nanoparticles in a cold, highly dilute gas (in the region of low vacuum pressures) above the sublimation interface, their self-organization into solid aggregates with a high specific surface area was observed; see Fig. 3.4.

The process of self-organization of nanoparticles at the sublimation phase interface was theoretically analyzed for different modes of vacuum sublimation of frozen liquid nanodispersion. At the same time, areas of different morphologies of particle nanostructures were also mapped by systematic experiments in terms of the influence of physical parameters on the process of controlled sublimation. They achieved high values of specific surface area (SSA) of lamellar nanoaggregates of the order of hundreds of m^2/g in comparison with units of $m^2 \ g^{-1}$ for the same particulate material obtained by the standard method of thermal drying.

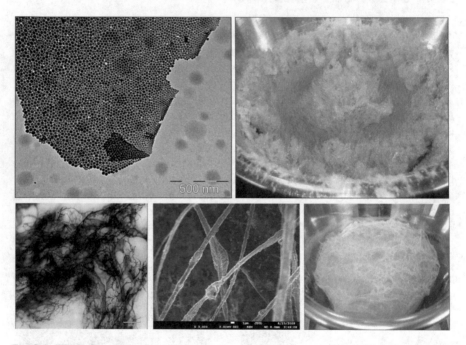

Fig. 3.4 TEM micrograph of lamellar aggregates of globular nanoparticles of fullerite nC_{60} with a diameter of about 26 nm by transmission electron microscopy (top left), macrophotography of the same fullerite nC_{60} aggregates in a metal sublimation dish (top right), micrograph of lamellar aggregates of silicon nanoparticles with a diameter of about 148 nm by confocal optical microscopy (bottom left), SEM micrograph of the same lamellar aggregates of silicon nanoparticles, which was rolled up into a relatively bulky layered fiber (bottom middle) and macrophotography of the same silicon aggregates in a metal sublimation dish (bottom right)

Based on the research performed so far, the self-organization of nanoparticles at the sublimation interface, a new technology for the preparation and modification of special nanomaterials with high values of the active surface was created. The basic parts of the technical solution are already protected by several patents [4]. This purely technical information is presented here mainly because in-depth patent searches in several countries also carried out a search of the literature, in which no relevant information in the relevant field was found. These would, of course, preclude the recognition of the novelty necessary for the grant of a patent. It is therefore a completely new issue, the technical application of which is patent-protected within a wide range of decisive parameters, but the relevant know-how for optimal applications is not formulated within the patent. Partly, the necessary data for its general formulation have already been obtained during the experiments, but due to the great diversity of dispersed materials, the physicochemical principles of the process of controlled sublimation need to be specified in more detail by deeper research.

3.2.1 Standard Freezing of Liquid Nanodispersion and Segregation of the Dispersion Fraction

One of the basic problems of self-organization of nanoparticles into lamellar structures in the above-mentioned process of controlled sublimation is to maintain an even distribution of the particles of the dispersion fraction in the liquid dispersion medium, which in most cases is formed by water. In order to sufficiently preserve the homogeneous structure of the spatial distribution during solidification, segregation changes of the local particle concentration must be avoided. These lead to final states with a significantly different morphology, where in particular the formation of fibers in the intercrystalline regions of increased concentration of segregated particles [5–7]. In order to maintain the homogeneity of the spatial distribution of particles from the liquid state to the solid state, it is necessary to suppress all the effects of segregation redistribution during slow crystallization. This is achieved by applying a very fast solidification process when freezing the liquid dispersion.

3.2.2 Flash Freezing of Liquid Nanodispersion

To ensure optimal conditions for the self-organization of lamellar nanoaggregates, the solidification of the aqueous dispersion is usually performed by variants of the flash freezing method. In the pure application of the flash freezing method in obtaining a highly amorphous form of ice, the fundamental condition is highly pure demineralized water without any microparticles. If the liquid is at rest without shocks, it can be supercooled to a temperature of approx. $-43\,°C$ [8]. The state transformation, caused for example by a blow, then takes place at an extreme rate throughout the volume, so that nucleation and crystal growth are suppressed and dense amorphous ice is formed. However, in an aqueous dispersion, heterogeneous nucleation of crystalline nuclei will most likely occur on the phase surface of the dispersed particles, and this method is practically unusable for its amorphous solidification.

A relatively simple alternative in the case of nanodispersions is the preparation of so-called LDA amorphous ice with low density (low-density amorphous ice), which is formed by flash solidification (approx. 10^5 K s^{-1}) of aerosol microdroplets on a smooth surface with a temperature below -50 °C. In this case, there is a classical mechanism of temperature drop by conducting heat through the phase interface on the cooling surface, which is very effective for fast freezing due to the small volume.

A technologically demanding method of global volume solidification of aqueous nanodispersion is its cooling to the water–ice interface in the state diagram at an increased pressure of the order of hundreds of MPa (this is done in a high-pressure chamber by using a high-pressure multiplier up to 420 MPa [9]). After a sudden release, a very rapid drop in pressure in the entire volume (with respect to the high speed of sound almost simultaneously) causes a transition to the solid region, with the formation of a glassy state of ice (see Fig. 3.5).

In many cases, however, the effect of segregation is not very pronounced, and for solidification, it is possible to use rapid freezing of liquid homogenized water at 0 °C in a deep-freezing box below -50 °C. Thus, there are several ways to rapidly solidify an aqueous dispersion of nanoparticles, which is further subjected to controlled sublimation in a vacuum recipient.

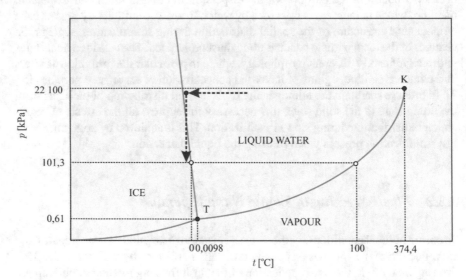

Fig. 3.5 Approximate scheme of pressure decompression (red phase trajectory) during solidification of high-pressure water, cooled to a temperature just above the phase curve between ice and water [10]

3.2.3 Influence of Temperature and Pressure on the Rate of Retraction of the Sublimation Interface During Evaporation of Frozen Liquid in Vacuum

In the process of controlled sublimation, the temperature and pressure at the sublimation phase interface are fixed by heating and sucking air through a needle microvalve at selected values, measured by a Pirani vacuum gauge behind the freezing chamber of the lyophilizer. This sets the sublimation flow of water molecules into the free space of the vacuum recipient and thus the rate of releasing of particles, homogeneously distributed in ice; see Fig. 3.6.

The mutual interaction of dispersion nanoparticles at the free evaporation interface is significantly different in cases of evaporation of water molecules from liquid and solid (frozen) states. This significantly affects the character of the resulting dry powder. The elementary conditions that have a fundamental influence on the formation and stability of the final bond between two nanoparticles can be characterized by the following factors:

(A) Mutual velocity of particles at the moment of collision
(B) Binding energy between aggregated particles
(C) Surface effects of interfacial tension and ζ-potential in the immediate vicinity of particles

(ad A) The magnitude of the relative velocity of the two particles in the liquid state of the dense dispersion is determined by the instantaneous velocity of the chaotic

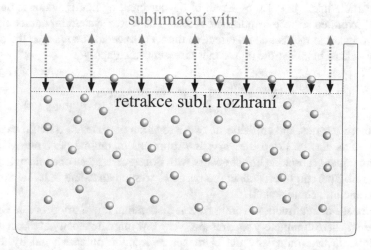

Fig. 3.6 Schematic representation of the retraction of the sublimation interface, caused by the evaporation of water molecules into the space of the vacuum recipient. At higher densities, evaporation manifests itself as a directed molecular flux, which we have laboriously called the "sublimation wind."

Brownian motion of the individual particle. At present, it is already possible to determine this microscopic characteristic by measurement and it can be considered as a real physical parameter. The instantaneous velocity of globular silicon microparticles of 3 μm, measured in the publication of Li et al. [11] reached an effective size of $v_{rms} = 0.422$ mm s^{-1} and with a low frequency of extreme values below 2 mm s^{-1}. Due to the fact that the interaction potential between nanoparticles has the character of superposition of very weak van der Waals interactions of individual atoms, in the case of the liquid state of the dispersion there is a considerable predominance of kinetic energy. For model of globular silicon microparticle with a radius $r = 1.5$ μm, at normal pressure and temperature 297 K, the effective kinetic energy of the particle itself reaches a value

$$E_k = \frac{1}{2}\left(\rho_{Si}\frac{4}{3}\pi r^3\right)v_{mrs}^2 \approx 18.3 \text{ meV}. \tag{3.1}$$

In conventional drying of a wet filter cake at a temperature above 60 °C, the mean kinetic energy of the same particle corresponds to the Brownian motion in a dense but liquid aqueous dispersion. Based on experimental analysis of the speed of Brownian particles in works [11, 12], the equipartition theorem in hydrodynamic modification for a liquid medium can be used to estimate the mean kinetic energy in a liquid.

$$E_k^{eqp} = \frac{3}{2}kT = 38.4 \text{ meV} \tag{3.2}$$

In Brownian motion of particles in a liquid, the kinetic energy is carried not only by the particle itself, but also by a part of the surrounding liquid, which is bound to the often hydrated surface by adhesive forces. The whole system is characterized by the hydrodynamic radius and its effective mass m^* exceeds the mass of the particle itself by at least half. For the above measurements then apply

$$E_k^{eqp} = \frac{m^*}{m}E_k < E_k \rightarrow m^* \approx 2\,m. \tag{3.3}$$

For these reasons, the collision of two Brownian particles in a liquid cannot be considered as elastic. From the very first moments of mutual penetration of both liquid containers, a braking effect occurs with different degrees of exchange of their momentum. This can often deflect the particles to such an extent that their surfaces do not come into contact at all.

In the case of sublimation from a frozen dispersion, it is a dry process. Brown's movement is here completely suppressed. Very low mutual velocity at direct contact of particles is determined by the Hamaker interaction potential and by the rate of agglomeration of gradually released particles in the surface layer above the sublimation interface; see Fig. 3.7.

Unlike liquid dispersion, the frequency of collisions in this mode is significantly lower. When a bound state of two or more particles is formed, the probability of its

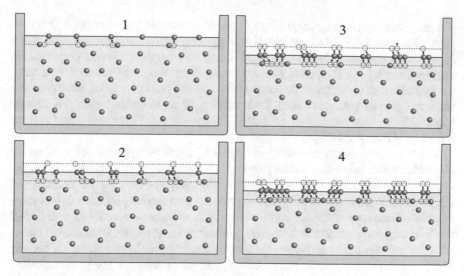

Fig. 3.7 Scheme of gradual accumulation of particles released at the sublimation interface during its retraction due to evaporation of water into vacuum

disruption by another collision is much lower than in the liquid dispersion. In the case of a very dense liquid dispersion, the frequency of mutual contacts of Brownian particles is high. The tight arrangement and the very numerous impacts also provide the conditions for the optimal orientation of the contact surfaces in large areas. In standard thermal drying, dense aggregates with a lower specific surface area are preferably formed.

(ad B) At the value of the specific surface energy of silicon (3.12) $A_{Si} = 1240 \, mJ \, m^{-2}$ [13] and assuming a perfectly oriented contact of two globular particles on 1‰ surface ($\kappa = 10^{-3}$), the estimate of the bond energy is

$$\Delta E^{111} = A_{Si}^{111} \kappa \pi d^2 \approx 54 \, keV. \tag{3.4}$$

Thus, in both drying processes, the binding energy is dominant compared to the kinetic energy. However, there is a very significant difference in the frequency of mutual contact of the particles. The bond, characterized by an energy of about 54 keV according to relation (3.4) is, with regard to the strong condition of mutual orientation, realized with a very low probability of the order of 10^{-6}. However, in a dense wet filter cake, this low probability will be compensated by the relatively high frequency of Brown particle collisions. During thermal drying, a tight and stable arrangement will prevail, in which the possible effect of self-organization at the microscopic level is suppressed. Such a thermally dried material is characterized by a dense structure with a low specific surface area.

(ad C) The formation and stability of the final bonds between the nanoparticles in a liquid medium are also significantly affected by surface effects, such as surface tension at the free surface and interfacial tension at the vessel walls and surface charge, given by the magnitude of the ζ-potential. Outside the interval $(-30\ \text{mV}; +30\ \text{mV})$ in most cases, the ζ-potential prevents aggregation of particles in the liquid by Coulomb repulsion. In systems with lower absolute magnitude of ζ-potential, the aggregation plays an important role and during the evaporation of water molecules, the particles remain in close proximity. Thus, in the final phase, the activity of the surface atoms of the material in the formation of more stable bonds can also be applied. In the dry process of controlled sublimation, the ζ-potential and surface effects (ad C) do not apply.

The basis for the preparation of lamellar nanostructures by the method of controlled sublimation in vacuum freezing is the minimization of the kinetic energy at the contact of particles (ad A) in favor of the energy of their mutual bonding (ad B). The maximum particles convergence velocity is essentially equal to the very low velocity of decrease of the sublimation interface. Its value is practically zero in real conditions of vacuum sublimation. A significant influence on the formation of bonds between particles on the surface of the sublimation interface has the speed and density of the output water vapor flow, which at higher values creates a phenomenon called "sublimation wind." For the saturated vapor pressure of water p_{sat} at temperature T, a semiempirical formula was derived based on thermodynamic analysis and a large number of experiments [14]

$$p_{\text{sat}}(T) = p_t \exp\left[\sum_{i=1}^{3} a_i \left(\frac{T}{T_t}\right)^{b_i-1}\right] \left\{ \begin{array}{lll} a_1 = -21.214; & a_2 = 27.320; & a_3 = -6.106 \\ b_1 = 0.0033; & b_2 = 1.207; & b_3 = 1.703 \\ p_t = 611.657\ \text{Pa}; & T_t = 273.16\ \text{K} \end{array} \right\}. \quad (3.5)$$

Based on this formula, the value of the maximum particle convergence velocity of two nanoparticles was derived. For the purpose of this derivation, we introduce a multiplication parameter $k < 1$, which quantifies the relationship between the partial vapor pressure of water $p_{\text{vac}}(T)$ in vacuum at temperature T and the corresponding saturated vapor pressure $p_{\text{sat}}(T)$

$$p_{\text{vac}}(T) = k \cdot p_{\text{sat}}(T). \quad (3.6)$$

Under the conditions given by Eq. (3.6), the partial vapor pressure of water $p_{\text{vac}}(T)$ in vacuum is lower than the saturated vapor pressure of water, and at the sublimation interface, there is an imbalance between the number of water molecules n_- emitted to vacuum and the number of n_+ molecules absorbed from "vakuum" back to volume of ice on area S for the period τ.

$$n_- = N_A \frac{p_{\text{sat}}(T)}{\sqrt{2\pi M_{\text{H}_2\text{O}} R T}} S\tau, n_+$$

$$= N_A \frac{p_{\text{vac}}(T)}{\sqrt{2\pi M_{\text{H}_2\text{O}} R T}} S\tau \left\{ \begin{array}{l} M_{\text{H}_2\text{O}} = 0.018\ \text{k gmol}^{-1}, \\ R = 8.314\ \text{J K}^{-1}\ \text{mol}^{-1}, \\ N_A = 6.022 \times 10^{23}\ \text{mol}^{-1} \end{array} \right\} \quad (3.7)$$

Their difference Δn represents the number of water molecules emitted into the vacuum by the area S of ice during the period τ.

$$\Delta n = n_- - n_+ = N_A \frac{p_{sat}(T)(1-k)}{\sqrt{2\pi M_{H_2O}RT}} S\tau \tag{3.8}$$

If $\overline{\rho}_o$ is the average ice density in the appropriate temperature range and the retraction speed of the sublimation interface is v_o, then for the sublimation loss Δm the weight of ice on the area S for the period τ applies

$$\Delta m = \overline{\rho}_o S v_o \tau. \tag{3.9}$$

Both parameters Δn and Δm in Eqs. (3.8) and (3.9) are unambiguously linked and provide a formula for the retraction speed of the sublimation interface

$$v_o(T) = \frac{p_{sat}(T)(1-k)}{\overline{\rho}_o} \sqrt{\frac{M_{H_2O}}{2\pi RT}}. \tag{3.10}$$

After the substitution by the semiempirical Formula (3.5), the retraction speed of the sublimation interface has final formula

$$v_o(T) = \frac{1}{\overline{\rho}_o} \sqrt{\frac{M_{H_2O}}{2\pi RT}} p_t (1-k) \exp\left[\sum_{i=1}^{3} a_i \left(\frac{T}{T_t}\right)^{b_i-1}\right] \quad \begin{array}{ll} a_1 = -21.214 & b_1 = 0.0033 \\ a_2 = 27.320 & b_2 = 1.207 \\ a_3 = -6.106 & b_3 = 1.703 \end{array}$$

$$\left\{ \begin{array}{l} p_t = 611.657 \text{ Pa}; \quad T_t = 273.16 \text{ K}; \\ \overline{\rho}_o = (926 \pm 0.005) \text{ kg m}^{-3}; M_{H_2O} = 0.018 \text{ kg mol}^{-1} \end{array} \right\}. \tag{3.11}$$

Figure 3.8 shows the dependence of the retraction speed of the sublimation interface $v_0(t)$ at a vacuum depth corresponding to 96% of the saturated vapor pressure of water at temperature t.

The speed of sublimation of the pure ice under well-defined physical conditions was measured in a vacuum recipient (20 l) above the freezing chamber of lyophilizer $-60\,°C$ with a 12.6 cm^2 of its vacuum inlet opening. The experiment was performed at a pressure in the vacuum recipient $p \approx 12.3$ Pa and at sublimation surface temperature $-40\,°C$. In the given experimental setup, the time dependence of the weight loss of ice Δm from the free sublimation surface area $S_{sub} = 531$ mm^2 inside the cylindrical vessel was observed. This dependence is illustrated by the graph in Fig. 3.9.

At a sublimation surface temperature of $-40\,°C$, an equilibrium of sublimation and desublimation flow would be achieved at a pressure equal to the saturated vapor pressure of water $p_{sat}\,(-40\,°C) = 12.8$ Pa. The pressure difference of 0.5 Pa corresponding to approx. 4% guarantees a defined sublimation flow, which is a basic condition for the application of sublimation control technology. Linear regression on the experimental data in Fig. 3.9 corresponds to equation

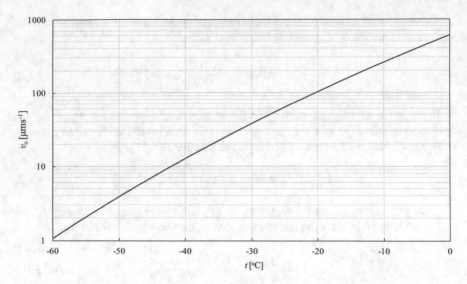

Fig. 3.8 Dependence of the retraction speed of the sublimation interface at the vacuum depth corresponding to 96% of the saturated vapor pressure of water at temperature t [15]

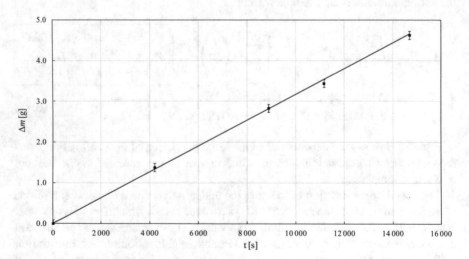

Fig. 3.9 Time dependence of sublimation weight loss of ice on an area of 531 mm^2 at sublimation surface temperature of $-40\,°C$

$$\Delta m = \left(\overline{\rho}_o v_{o\mathrm{exp}} S_{\mathrm{sub}}\right) \cdot t. \tag{3.12}$$

It follows from its direction that under the given conditions at the temperature of the sublimation surface $-40\,°C$ and the pressure of 12 Pa the retraction speed of the sublimation interface reaches the value v_{exp} (233.15 K) $= 0.61\ \mu\mathrm{m\ s}^{-1}$. During the sublimation of water molecules from a frozen nanodispersion, most nanoparticles bind to the surface of ice–vacuum sublimation interface. By its sublimation retraction, the released particles are entrained by the volume of ice toward the other particles of the dispersion not yet evaporated (see Fig. 3.6) and the maximum velocity of their mutual convergence is equal to the above-mentioned retraction speed v_0 (T). In the case of mutual contact of two released particles above the sublimation interface, their very low relative velocity is additionally supplemented by a component which is given to it by the weak attraction by the action of the Hamaker potential.

In a very sparse arrangement during controlled vacuum freezing, the bond will be usually formed at the first contact of the particles without the disturbing interactions of other impacts, such as in a liquid. In this situation, the undisturbed effect of self-organization is much more pronounced, which leads to the formation of larger lamellar aggregates above the surface of the sublimation interface (see Fig. 3.7).

3.2.4 Vapor Pressure Gradient Above the Sublimation Interface and "Sublimation Wind"

The situation when the pressure p in the vacuum recipient is equal to the saturated vapor pressure p_{sat} (T) at a given temperature T, corresponds to a state of sublimation equilibrium in which the weight of ice in the sublimation vessel does not change—sublimation and desublimation flow are mutually compensated. For the effective implementation of the controlled sublimation technology, it is necessary that the pumping limit of the vacuum pump is as low as possible below the level of saturated vapor pressure of water, the dependence of which on the temperature $p_{sat}(t\ °C)$ of the sublimation surface is shown in the graph on Fig. 3.10.

The intensity of the sublimation flow can then be regulated both by the temperature t of the phase interface by means of infrared heating and by the pressure difference $\Delta p(T) = p_{sat}(T) - p$ by means of vacuum fixation.

Between the sublimation interface and the outlet level of the vessel in Fig. 3.11, due to the pressure difference $\Delta p(T)$ and the expansion of water vapor, the flow into the vacuum recipient occurs. Under conditions where the output length is not significantly longer than the diameter of the sublimation interface, we will consider the pressure gradient to be constant over the entire length x, and the vapor pressure at a very small height h above the sublimation interface can be expressed by a linear equation

$$p_h = \nabla p \cdot h = \frac{p_{sat}(T) - p}{x}h. \tag{3.13}$$

If we introduce a dimensionless parameter $\lambda = p/p_{sat} \leq 1$ to express the pressure p in a vacuum recipient, then the linear dependence (3.13) takes the form suitable for later simplification.

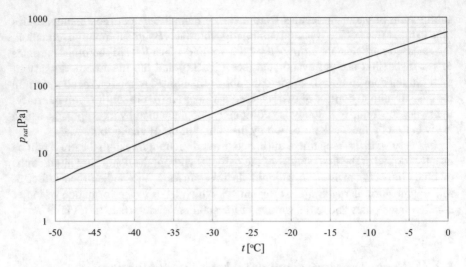

Fig. 3.10 Dependence of saturated vapor pressure of water on temperature in °C [14]

Fig. 3.11 Schema of a cylindrical sublimation vessel with an indication of the outlet length x between the sublimation interface and the "practically infinite" volume of the vacuum recipient, and a transition layer of thickness h in which the interaction between the particles and the process of their self-organization takes place immediately

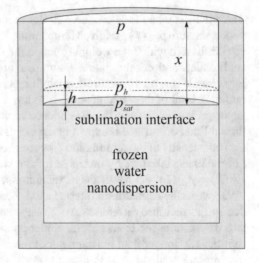

$$p_h = p_{sat}(T)\frac{h}{x}(1 - \lambda). \qquad (3.14)$$

Similar to the pressure in Fig. 3.10, the dependence of density of the saturated vapor pressure of water on the temperature, Fig. 3.12.

The expansion of highly dilute water vapor in the immediate vicinity above the sublimation interface will be approximated by the Boyle–Mariott equation for the behavior of an ideal gas (the temperature change can be neglected in a very thin layer of height h). Using the formulation of the above Boyle–Mariott equation for vapor

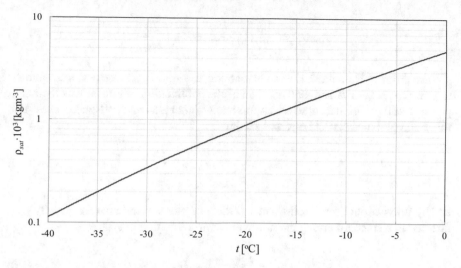

Fig. 3.12 Dependence of density of the saturated vapor pressure of water on the temperature °C [14]

density, its value ρ_h at height h above the sublimation interface can be derived from Eq. (3.14).

$$\frac{\rho_h}{p_h} = \frac{\rho_{sat}(T)}{p_{sat}(T)} \rightarrow \rho_h = \rho_{sat}(T) \cdot \frac{h}{x}(1 - \lambda). \tag{3.15}$$

For the mean square velocity of water molecules emitted from the sublimation interface at temperature T, the Maxwell–Boltzmann distribution within the kinetic theory gives the value

$$\overline{v}^2 = \frac{2RT}{M_{H_2O}}. \tag{3.16}$$

While in the immediate vicinity of the sublimation interface a model of the isotropic motion of sublimated molecules can be accepted as in a spatially limited saturated vapor pressure of water, only the components of quadratic velocities perpendicular to the interface need be taken into account to describe the sublimation wind into vacuum.

$$\overline{v}_\perp^2 = \frac{1}{4\pi} \int_0^{\frac{\pi}{2}} \int_0^{2\pi} \left(\overline{v}^2 \cos\theta\right) \sin\theta \, d\theta \, d\varphi \rightarrow \overline{v}_\perp^2 = \frac{1}{4}\overline{v}^2. \tag{3.17}$$

By the simultaneous application of Formulas (3.15), (3.17) and (3.18), the flow of the sublimation wind can be expressed in the form $J_{sub} = \rho_{sat}\overline{v}_\perp$,

$$J_{\text{sub}}(h) = \frac{1}{2}\frac{h}{x}(1-\lambda)\rho_{\text{sat}}(T)\sqrt{\frac{2RT}{M_{H_2O}}}. \tag{3.18}$$

In addition to Hamaker's mutual interaction, external forces also play an important role in the process of self-organization at the sublimation interface. It is the rising pressure of the sublimation wind, approximately characterized by Newton's resistive force of a globular particle of cross section S

$$F_N = \frac{1}{2}\underbrace{C}_{0.5}\rho_h S \bar{v}_{\perp}^2 \tag{3.19}$$

and by gravitational force acting on a particle of radius r and material density ρ_{part} directed in a standard arrangement toward the sublimation interface

$$G = \left(\frac{4}{3}\pi r^3 \rho_{\text{part}}\right)g. \tag{3.20}$$

In a situation where the resistance force of the sublimation wind lift exceeds the gravitational force, the particles released from the sublimation interface are entrained in the vacuum recipient and mutual bonds and self-organization do not occur. The applicability of controlled sublimation technology is therefore limited by the boundary condition of force balance $F_N \leq G$,

$$\frac{1}{4}\rho_{\text{sat}}(T) \cdot \frac{h}{x}(1-\lambda)\pi r^2 \frac{1}{4}\frac{2RT}{M_{H_2O}} \leq \frac{4}{3}\pi r^3 \rho_{\text{part}}g. \tag{3.21}$$

This condition limits the radius r of the dispersed particles to a minimum value

$$r \geq \frac{3}{32}\left(\frac{h}{x}\right)\frac{(1-\lambda)R}{\rho_{\text{part}}M_{H_2O}g}T\rho_{\text{sat}}(T). \tag{3.22}$$

The application limit defined in this way can be regulated in controlled sublimation by setting decisive physical parameters such as the definition of the pressure p in the recipient using the parameter $\lambda = p/p_{\text{sat}} \leq 1$, the temperature T of the sublimation interface and the depth x of the sublimation vessel. Figure 3.13 graphically shows the dependences of the limit equilibrium radii of the particles on the pressure and depth of the sublimation vessel for the values $\lambda = 0.96$ and $x = 1, 2, 3, 4$ and 5 cm. The thickness of the sublimation flow calculation layer $h \approx 1$ Å was minimized so that the output sublimation flow flowed around most of the volume of released particles at the sublimation interface.

Especially to the last parameter x, it should be noted that with its growth, the dynamic effect of the sublimation wind gradually decreases and within the limit of infinite length, the movement of water molecules has an isotropic character associated with the state of sublimation equilibrium. This situation can be approached by using

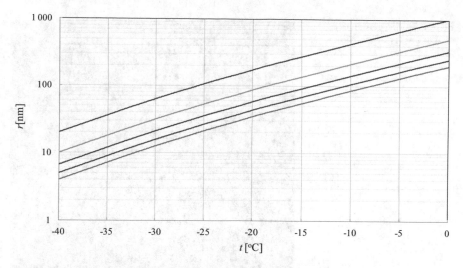

Fig. 3.13 Temperature dependences of "equilibrium radii" of particles in the elevation of the molecular flow of the sublimation wind. Dependencies were calculated for the region just above the sublimation interface for the value of the parameter $h = 10$ nm. The curves are color-coded for different values of depth x: — 1 cm, — 2 cm, — 3 cm, — 4 cm, — 5 cm

the Knudsen effusion cell. Mentioned device is mostly used to generate a collinear molecular beam, which is emitted from an effusion orifice with a diameter that is much smaller than the mean free path of saturated vapors inside the cell. When applied to controlled sublimation, this design ensures extremely low sublimation intensity and the released particles are not destabilized by sublimation wind pressure during mutual interactions.

This can result in significantly more interconnected 3D network aggregates, as shown in Fig. 3.14. In the case of an open sublimation vessel, on the other hand, the effect of the sublimation wind on the self-organization is destructive. In the case of aggregation growth of a particulate 2D monolayer over the area of the sublimation interface, as shown in Fig. 3.15, the sublimation surface is gradually covered by a lamellar layer, under which the pressure of the expanding vapors subsequently increases. This results in a breach of the integrity of the lamellar layer until it is completely torn. This situation is illustrated by the photograph in Fig. 3.16, which captures torn lamellar aggregates of nC_{60} globular nanoparticles fluttering in the molecular current of a sublimation wind.

Fig. 3.14 Micrograph of a
network 3D structure of
SiO$_2$ nanoparticles with a
size of about 8 nm
aggregated in an effusion cell

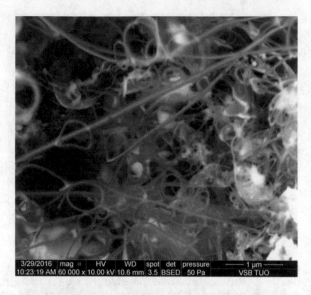

Fig. 3.15 Micrograph of a
lamellar 2D structure,
formed at the sublimation
interface by self-organization
of silicon nanoparticles, on
which the connection is
broken due to vapor pressure
and sublimation wind

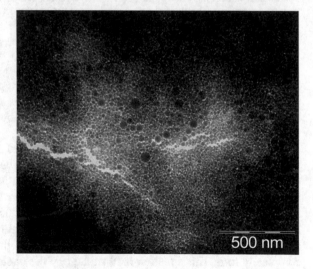

3.2.5 Influence of Nanoparticle Density in Dispersion on Aggregation Rate

The decisive factor for the speed and quality of the self-organization of the particles
at the sublimation interface is the mean frequency of the release of the particles from
the fixation from the frozen dispersion medium. This mean frequency is determined
by two basic factors:

1.   Sublimation flow, described by Eq. (3.18).

Fig. 3.16 Photograph of large lamellar nanoaggregates of fullerite globular nanoparticles nC$_{60}$ with a medium size of about 26 nm, fluttering in the molecular current of the sublimation wind

2. Volume concentration of the number of particles in the frozen dispersion.

At a constant concentration, the required mean frequency can be achieved by choosing the temperature and pressure, which define the appropriate size of the sublimation flow and thus the rate of retraction of the sublimation interface. On the other hand, at a constant value of the sublimation flow, the desired mean frequency can be achieved by choosing a suitable concentration of the number of particles in the frozen dispersion. In contrast to the first variant, in the second case it is possible to first optimize the sublimation flow with a view to minimizing its destructive effect, and then to select the particle concentration so as to achieve the desired mean frequency of releasing particle and the associated efficiencies in the preparation of aggregates.

3.3 Morphological Structure of Lamellar Aggregates of Nanoparticles

3.3.1 Statistical Size Distribution of Free Nanoparticles and Their Total Specific Surface Area SSA

The aim of the following analysis is to compare the total specific surface area Σ of isolated particles in the liquid dispersion with the values for standard thermally dried material and material. For isolated globular particles, its value is determined on the basis of knowledge of their material density ρ_p and statistical volume distribution P_v (d) [16].

For a statistical distribution P_v (d) of the total volume of globular silica particles in Fig. 3.17, the total specific surface area Σ can be expressed using Formula [16]

$$\Sigma = \frac{6}{\rho_p} \sum_{i=1}^{\infty} \frac{P_v(d_i)}{d_i}. \tag{3.23}$$

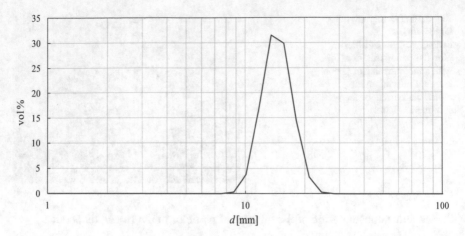

Fig. 3.17 Statistical distribution of the total volume of silica particles according to the size of the hydrodynamic diameter was measured by the dynamic light scattering method on a Malvern Zetasizer Nano ZS instrument at a laser wavelength of 663 nm. The mean particle size corresponds to 15 nm

Its size represents the sum of the surfaces of all isolated particles per 1 g of material, and in the case of the above-mentioned silica nanoparticles, it was 181 m^2g^{-1}. After drying the particulate dispersion by the classical method of thermal evaporation from the liquid state, a value of only 4 $m^2\ g^{-1}$ was measured for the total specific surface area Σ by the BET method; see Fig. 3.16.

The very low value of the specific surface area of the powder $\Sigma = 4\ m^2\ g^{-1}$ (Fig. 3.18) is only 2.2% of the surface area of the insulated particles according to (3.23). This fact indicates a significant ratio of contact bonding areas between the particles, which reduces the overall specific surface area of the material.

3.3.2   Shielding Effect in Lamellar Aggregates of Nanoparticles After Controlled Sublimation and Its Effect on SSA

Na rozdíl od standardních metod termálního sušení kapalinových disperzí vykazují agregované nanostruktury po řízené sublimaci velmi malý podíl kontaktních vazebných ploch, které snižují celkový specifický povrch materiálu. In contrast to standard methods of thermal drying of liquid dispersions, aggregated nanostructures after controlled sublimation show a very small ratio of contact bonding surfaces, which reduce the overall specific surface area of the material.

For the above-mentioned silica nanoparticles, dried by controlled sublimation, the measured value is about 154 $m^2\ g^{-1}$ (Fig. 3.19), which corresponds to 85% of the free surface of the isolated particles according to (3.23). The area enclosed by

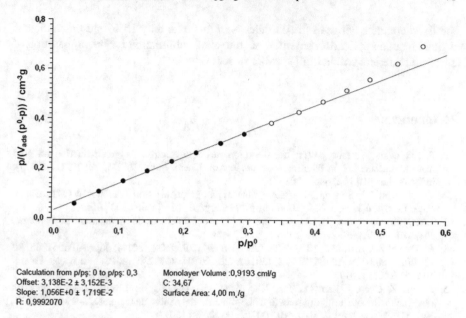

Calculation from p/pş: 0 to p/pş: 0,3
Offset: 3,138E-2 ± 3,152E-3
Slope: 1,056E+0 ± 1,719E-2
R: 0,9992070

Monolayer Volume :0,9193 cml/g
C: 34,67
Surface Area: 4,00 m,/g

Fig. 3.18 Recording of the measurement result of the specific surface of thermally dried silica powder by the BET analysis method

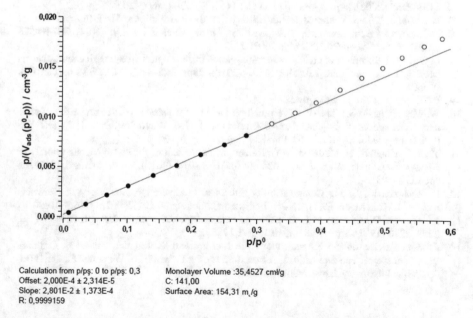

Calculation from p/pş: 0 to p/pş: 0,3
Offset: 2,000E-4 ± 2,314E-5
Slope: 2,801E-2 ± 1,373E-4
R: 0,9999159

Monolayer Volume :35,4527 cml/g
C: 141,00
Surface Area: 154,31 m,/g

Fig. 3.19 Recording the result of specific surface measurements by BET analysis for silica nanoparticles, dried in a vacuum by controlled sublimation technology

the fixed contacts between the particles is in this case only 15%, and this indicates a considerable application potential of controlled sublimation for the preparation of sorption materials with a high specific surface area.

References

1. F. Lu, et al., Surface proton transfer promotes four-electron oxygen reduction on gold nanocrystal surfaces in alkaline solution. J. Am. Chem. Soc. **139**(21), 7310–7317 (2017). https://doi.org/10.1021/jacs.7b01735
2. P.R. Somani, S.P. Somani, M. Umeno, Planer nano-graphenes from camphor by CVD. Chem. Phys. Lett. **430**(1–3), 56–59 (2006). https://doi.org/10.1016/j.cplett.2006.06.081
3. D.R. Huffman, Synthesis, structure, and properties of fullerenes, in *Nanomaterials: Synthesis, Properties and Applications* (1996), pp. 477–494
4. R. Dvorský, Czech-303513 (2012), USA-US 9,410,739 B2 (2016), Japan-JP5961266 B2 (2016), China-CN 103842755 B (2016), Russia-RU 2599282 (2016), European Patent-EP2751508B1 (2017)
5. J. Yan, Z. Chen, J. Jiang, L. Tan, X.C. Zeng, Free-standing all-nanoparticle thin fibers: a novel nanostructure bridging zero- and one-dimensional nanoscale features. Adv. Mater. **21**(3), 314–319 (2009). https://doi.org/10.1002/adma.200801130
6. T. Fukasawa, Z.-Y. Deng, M. Ando, T. Ohji, S. Kanzaki, Synthesis of porous silicon nitride with unidirectionally aligned channels using freeze-drying process, J. Am. Ceram. Soc. **85**(9), 2151–2155 (2002). https://doi.org/10.1111/j.1151-2916.2002.tb00426.x
7. H. Zhang, J.-Y. Lee, A. Ahmed, I. Hussain, A.I. Cooper, Freeze-align and heat-fuse: microwires and networks from nanoparticle suspensions. Angew. Chemie Int. Ed. **47**(24), 4573–4576 (2008). https://doi.org/10.1002/anie.200705512
8. C. Goy, et al. Shrinking of rapidly evaporating water microdroplets reveals their extreme super-cooling. Phys. Rev. Lett. **120**(1), 015501 (2018). https://doi.org/10.1103/PhysRevLett.120.015501
9. PTV, spol. s r.o. http://www.ptv.cz
10. W. Wagner, A. Pruß, The IAPWS formulation 1995 for the thermodynamic properties of ordinary water substance for general and scientific use. J. Phys. Chem. Ref. Data **31**(2), 387–535 (2002). https://doi.org/10.1063/1.1461829
11. T. Li, S. Kheifets, D. Medellin, M.G. Raizen, Measurement of the instantaneous velocity of a Brownian particle. Science (80-) **328**(5986), 1673–1675 (2010). https://doi.org/10.1126/science.1189403
12. R. Zwanzig, M. Bixon, Compressibility effects in the hydrodynamic theory of Brownian motion. J. Fluid Mech. **69**(1), 21–25 (1975). https://doi.org/10.1017/S0022112075001280
13. J.J. Gilman, Direct measurements of the surface energies of crystals. J. Appl. Phys. **31**(12), 2208–2218 (1960). https://doi.org/10.1063/1.1735524
14. Revised Release on the Pressure along the Melting and Sublimation Curves of Ordinary Water Substance. The International Association for the Properties of Water and Steam, 2011, Available: http://www.iapws.org/relguide/MeltSub2011.pdf

15. R. Dvorsky, J. Trojkova, J. Lunaček, K. Piksova, O. Černohorsky, Synthesis of inorganic nanofibers and lamellar structures with large specific surface by means of controlled vacuum freeze-drying process, in *NANOCON 2011*, pp. 58–63 (2011)
16. R. Dvorsky, et al., Synthesis of core-shell nanoparticles Si-ZnS by reactive deposition of photo-catalytic ZnS layer on the surface of carrier Si nanoparticles in aerosol microdrops. Procedia Soc. Behav. Sci. **195**, 2122–2129 (2015). https://doi.org/10.1016/j.sbspro.2015.06.254

Chapter 4
Lamelar Aggregation of Nanoparticles From a Frozen Liquid at the Sublimation Interface—Mathematical Modeling

4.1 Experimental Basis for Mathematical Modeling of Self-organization of Globular Nanoparticles

4.1.1 Choice of Nanomaterial for Mathematical Modeling and Hamaker's Theory of Interaction Between Two Globular Nanoparticles of Fullerite nC_{60}

The integral form of the van der Waals interaction, which is described by the Hamaker potential, plays a dominant role in the mutual interactions of submicroparticles and nanoparticles released by the sublimation of water at the phase interface. Some of the experiments that used controlled sublimation technology have been described in our previous works [1–3]. Materials were selected for the study, for which the values of the Hamaker constant were available, necessary for further theoretical study of the process of self-organization. At the same time, the application potential for the preparation of materials with high specific surface area values was taken into account (Fig. 3.4). Such nanomaterials are suitable for sorption and catalytic processes. The technology of their preparation is protected by patents in several countries (CZ, USA, JP, EU, China and Russia) [4]. We state this fact mainly for the reason that the methodology of international patent research also guarantees the novelty of the mentioned issue and its technical solution. However, for a very specific formulation of technological know-how, a deeper understanding of the processes of self-organization is necessary, which will provide algorithms for the optimal setting of physical processes. At the same time, its mathematical modeling also proves to be a very powerful tool for studying the process of self-organization [5].

© The Author(s), under exclusive license to Springer Nature Switzerland AG 2022
R. Dvorsky et al., *Nanoparticles' Preparation, Properties, Interactions and Self-Organization*, SpringerBriefs in Applied Sciences and Technology,
https://doi.org/10.1007/978-3-030-89144-2_4

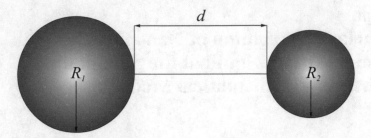

Fig. 4.1 Interaction scheme of geometric arrangement of two spherical nanoparticles of identical material with radii R_1 and R_2

Based on Hamaker's theory, the interaction potential was derived for two spherical nanoparticles with radii R_1 and R_2 (Fig. 4.1).

Within Hamaker's theory, the interaction potential of two spherical particles (see Fig. 4.1) from different materials 1, 2 is described in by Formula (2.20). The Hamaker constant A_{12} represents here a material parameter, the magnitude of which is characterized by a characteristic value of the order of 10^{-20} J. The theoretical calculation of its value of the Hamaker constant is based on Lifshitz's theory of van der Waals interactions [6–8]. However, the specific value is usually determined by experiments based on the interaction of the particulate material with electromagnetic radiation

$$A_{ij} = \left[\begin{array}{c} \frac{3}{4}kT\left(\frac{\varepsilon_{ri}-1}{\varepsilon_{ri}+1}\right)\left(\frac{\varepsilon_{rj}-1}{\varepsilon_{rj}+1}\right) \\ + \frac{3h\nu_e}{8\sqrt{2}} \frac{\left(n_i^2-1\right)\left(n_j^2-1\right)}{\sqrt{\left(n_i^2+1\right)\left(n_j^2+1\right)}\left(\sqrt{n_i^2+1}+\sqrt{n_j^2+1}\right)} \end{array} \right]. \tag{4.1}$$

The vacuum Hamaker constant A_{12} for the interaction of two different types of materials "I" and "j" in Eq. (4.1) is determined by absolute temperature T, relative permittivities ε_{r1} and ε_{r2}, refractive indices n_1 and n_2 for both materials, and absorption maximum frequency in ultraviolet area. For the interaction between particles of the same material, Eq. (1.3) can be simplified to the final Formula (4.2)

$$A_{ii} = \frac{3}{4}kT\left(\frac{\varepsilon_{ir}-1}{\varepsilon_{ir}+1}\right)^2 + \frac{3h\nu_e}{16\sqrt{2}} \frac{\left(n_i^2-1\right)^2}{\left(n_i^2+1\right)\sqrt{n_i^2+1}}. \tag{4.2}$$

Spherical fullerite nanoparticles with a mean size of 26 nm were chosen for model simulation experiments, see Fig. 3.4 on the left, the experimental preparation of which will be described below. In the case of fullerite, the relative permittivity ε_r is 3.5 and the refractive index n for 600 nm is 2.2 [9, 10]. The vacuum Hamaker constant, calculated on the basis of these parameters using Eq. (4.2) for the interaction between two fullerite spheres, is about $A_{nC60\text{-}nC60} = 7.8 \times 10^{-20}$ J and will serve as the basis for the construction of interaction potential for mathematical modeling of nC_{60} nanoparticle self-organization.

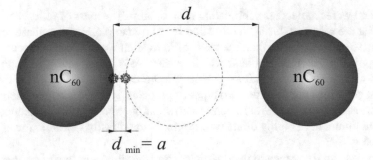

Fig. 4.2 Interaction distance d of the surfaces of fullerene nanoparticles nC_{60} reaches a minimum at a distance of one lattice constant of cubic fullerite $d_{min} = a$

4.1.2 Construction of the Interaction Potential for the Set of Globular Nanoparticles nC_{60}

For the model, a set of spherical nanoparticles nC_{60} [11] (Fig. 3.4 left) with mean radius $R = 13$ nm and the quantitative expression of the interaction potential by Eq. (2.20) with Hamaker constant about $A_{nC60 - nC60} = 7.8 \times 10^{-20}$ J is used. In mathematical modeling, the minimum distance d between the surfaces of two nC_{60} fullerite nanoparticles at the moment of contact must be determined in order to limit the potential for the stability of the numerical solution with (2.20). C_{60} fullerenes crystallize in the fcc lattice structure as fullerite crystals with a constant $a = 14.2$ Å [12]. This contact distance schematically shown in Fig. 4.2 will be considered as the minimum value of the parameter $d = d_{min}$.

Based on the physical analysis of the controlled sublimation process, a strategy for numerical simulation of nanoparticles aggregation at the phase interface at the Hamaker interaction potential (2.20) is further proposed.

4.1.3 Basic Conditions for Mathematical Modeling of Self-organization in the First 2D Approximation

In the first approximation, the process of self-organization was mathematically modeled for two dimensions of the sublimation interface based on the scheme in Fig. 3.6. During the evaporation of water molecules from the sublimation phase interface, the surface of the frozen liquid retracts, and the released dispersed particles gradually accumulate on it (Fig. 3.6). In the first 2D model approximation, particle dynamics is solved only with two (translational) degrees of freedom in the plane of the sublimation phase interface (rotational degrees of freedom were not considered in this first approximation). In this case, it is necessary to analyze the 2D system as a two-component mixture of fullerite nanoparticles with a radius of

13 nm and water molecules sublimating into vacuum. The mass of fullerite nanoparticles that are released during sublimation in the model plane of the phase interface is about 10^6 times greater than the mass of individual water vapor molecules. At a pressure in the vacuum recipient, which reaches 96% of the saturated vapor pressure of water and a temperature of -40 °C, we can assume a minimal dynamic effect in collisions between the two types of particles. It is necessary to analyze this assumption by numerical simulation of Brownian motion of fullerite nanoparticles in the environment of highly dilute water vapor before the final formulation of model algorithms.

The releasing frequency of nanoparticles from the surface of the frozen dispersion is another important parameter in 2D simulation. At the mean particles mass $m_{C_{60}} = 1.56 \times 10^{-20}$ kg and their mass fraction $x_{nC60} = 0.1$ wt%, the fullerite nanoparticles concentration C_{nC60} in the dispersion is given by the formula

$$C_{nC_{60}} = \frac{x_{nC_{60}} \rho_{H_2O}}{\left(1 - x_{nC_{60}}\right) m_{nC_{60}}} \quad \left(C_{nC_{60}} = 64 \times 10^{18} \text{ m}^{-3} = 64 \ \mu m^{-3}\right). \tag{4.3}$$

Since the 2D set at the sublimation interface has a grandcanonic character, the releasing frequency of dispersed nC_{60} nanoparticles from the unit area $\Phi_{nC_{60}} \left[\mu m^{-2} \text{ s}^{-1}\right]$ is important for its mathematical modeling and is equal to the product of the volume concentration C_{nC60} and the retraction rate of the sublimation interface $\upsilon_o = 0.61 \ \mu m \text{ s}^{-1}$ (see Fig. 3.7)

$$\Phi_{nC_{60}} = \upsilon_o C_{nC_{60}} \quad \left(\Phi_{nC_{60}} \doteq 39 \ \mu m^{-2} s^{-1}\right). \tag{4.4}$$

These parameters were used in the numerical simulation of fullerite aggregations of nC_{60} nanoparticles at the sublimation interface [3].

4.2 First Model Approximation—Planar Aggregation Model

4.2.1 Physical Conditions Over the Sublimation Interface and the Interaction Surface Layer Model

To formulate the first approximation of the 2D model, it is important to analyze the influence of water vapor molecules on the surface of nC_{60} nanoparticles. First, a simulation of Brownian motion of the isolated particle had to be performed. Regarding the mass ratio of water molecules and fullerite nanoparticles $(2.99 \times 10^{-26}$ kg)/$(1.56 \times 10^{-20}$ kg$) \approx 2 \times 10^{-6}$, we assume a negligible effect of Brownian motion. In this case, the nC_{60} particles will only move due to the interaction potentials (98).

Based on the kinetic theory of water molecules at -40 °C and 12.8 Pa, the mean frequency of impacts of water molecules on a spherical nanoparticle nC_{60} with a radius of 13 nm was calculated to 3.5×10^{14} s^{-1}. Due to the stochastic nature of Brownian motion, the simulation has been performed for six sets of twenty independent simulation experiments (Fig. 4.3).

To maximize the dynamic effect, the simulation was performed using only frontal elastic collisions between nanoparticles and water vapor molecules at a pressure of 96% saturated vapor pressure of water and a temperature of -40 °C. The motion

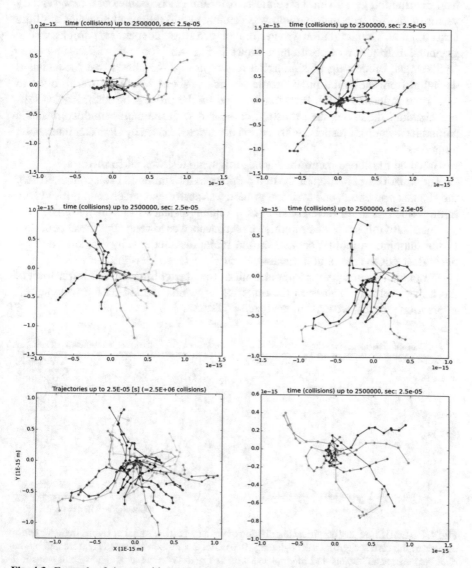

Fig. 4.3 Example of six sets with twenty variants of the Brownian trajectory of the nC$_{60}$ nanoparticle with zero initial velocity and initial position at the origin of the coordinate system $(x, y) = (0, 0)$

equation for calculating changes in the nC_{60} nanoparticle velocity at collision with a water molecule complies with the law of conservation of momentum and energy at each simulation step. The model time between two collisions was generated purely deterministically as the inverse value of their frequency, calculated from the kinetic theory of gases. The water molecule was generated at a random position on the surface of the nanoparticle with a radial velocity generated by a Maxwell–Boltzmann statistical distribution at a given temperature. After the elastic collision, a new velocity vector of the fullerite nanoparticle was calculated and remained constant until the next collision. The simulation was performed for a set of independent particles whose graphical distribution of positions is shown in Fig. 4.3.

Statistical processing of simulation results in Fig. 4.5 shows that the standard deviation from the initial nanoparticle position is about 4 Å in 5.4 μs. In order to assess the influence of the Brownian motion on the process of nanoparticles self-organization, the above data must be compared with the nanoparticles kinematic parameters which correspond with motion of particles caused by Hamaker interaction potentials.

Based on the above-mentioned needs following test was performed:

The equations of motion of the two particles in the simulation cell contained only the Hamaker potential, and their integration was performed under conditions of zero initial velocity and for different sizes of the initial distance r of both particles.

The aim of this series of simulation experiments was to determine the dependence of the duration to mutual contact on the initial distance r (Fig. 4.6 left), and the dependence of the maximum impact velocity on r (Fig. 4.6 right).

Simulation results give the upper limit of the impact time and the lower limit of the impact velocity. To assess the effect of Brownian motion on the nanoparticle interaction process, we need to analyze two factors.

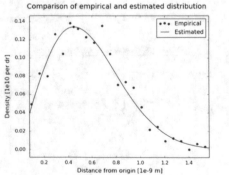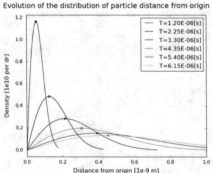

Fig. 4.4 (Left) Simulated empirical distribution of the nanoparticle's final positions frequency with respect to the distance r from the beginning (0, 0) after 2.5 million impacts is represented by the blue "experimental" points and interpolated with red regression curve of the Weibull statistical distribution. (Right) Scheme of time evolution (dissolution) of Weibull statistical distribution in the interval $1.20\ \mu s \le t \le 6.15\ \mu s$.

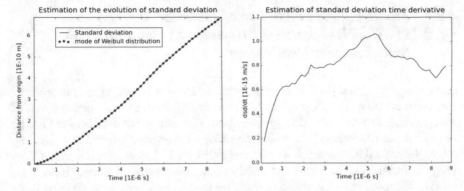

Fig. 4.5 (Left) Time dependence of the Weibull distribution statistical mode Fig. 4.4 shows extremely small movements of the nC$_{60}$ nanoparticles from their initial position $(x, y) = (0, 0)$ and (right) shows its time derivative, which appears to show an asymptotic plateau and justifies the possibility of linear extrapolation of the graph on the left

The first is the mean time between two creations of nanoparticles in a simulation cell (1×1) μm, which is 26 ms based on the calculation of the frequency (4.4) and which would correspond to a mean particle displacement of about 2 μm in the extrapolation of Brownian standard deviation according to the dependence on Fig. 4.5 (left).

The second is the impact time of two nanoparticles from their initial distance of 1 μm in the simulation cell, which is according to Fig. 4.6 (left) 16 μs and can be considered as a characteristic time of two nanoparticles interaction. Within this time, the Brownian standard deviation of the position is 12 Å. This maximum estimate gives a relative value of about 1‰ and shows that the Brownian motion has a negligible effect on the nanoparticles aggregation.

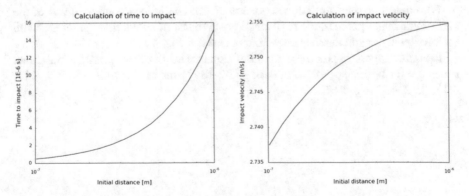

Fig. 4.6 Dependence of the time impact on the initial distance r between the particles (left) and the dependence of the impact velocity on the initial distance r between the particles (right)

4.2.2 Mathematical Simulation of Lamellar Aggregation of nC_{60} Nanoparticles Dispersion System with Known Statistical Size Distribution

After simplifying the process of self-organization by eliminating the influence of Brownian motion, we can proceed to a specific simulation of self-organization of nanoparticles in a vacuum 2D approximation. The final simulation is based on the particle dynamics method. It is an integration of Newton's equations of motion, where the mutual interaction is determined only by the Hamaker potential (2.20). Its program formulation has been extended to ensure numerical stability. To limit the relative distance below the collision limit, we used a quadratic elastic resistive force with equilibrium in the distance of the lattice constant of fullerite. The integration was performed by the fourth-order Runge–Kutta method. Particles are generated in the simulation cell in random positions with a uniform statistical distribution and mean frequency given by Eq. (4.4). Results of the analysis above show that the mean time between successive creations of two particles in the simulation cell is significantly higher than the maximum time of 16 µs to their contact. In this situation, the remaining time to create another particle would not be used efficiently and the computational time would be pointlessly extended. Therefore, when a bond is formed, the simulation time is shifted forward until a new particle is generated. Given the significant length of real time between two successive contacts of a cluster with a new particle, it can reasonably be expected that the 2D cluster has enough time in this "dead phase" to dynamically relax and achieve a stable configuration. For this reason, the time coordinate shift does not affect the actual process of creating the aggregated structure.

For the mathematical simulation of the self-organization process, we chose a system of globular fullerite nanoparticles with a mean size of 26 nm and statistical distribution frequency function according to the diameters in Fig. 4.7.

For control of the simulation sequence of five continuous configurations of the self-organization process, we performed positive test of simulation agreement with the predicted particle number growth in period 2 s, Fig. 4.8.

Bound configurations of the simulation sequence lasting up to 2 s are for individual times $t = \{128; 256, 4; 487, 2; 1026; 1538; 1897\}$ ms as shown in Fig. 4.9, 4.10, 4.11, 4.12, 4.13 and 4.14.

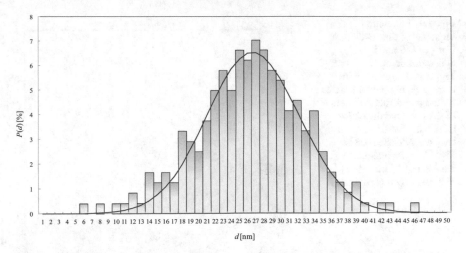

Fig. 4.7 Statistical distribution frequency function $P(d)$ of nC_{60} nanoparticles according to their diameter d with normal regression —— with a mean diameter of 26 nm. Image analysis of a representative selection of 240 nanoparticles of the TEM micrograph in Fig. 3.4 (left) then divided them into 50 size categories

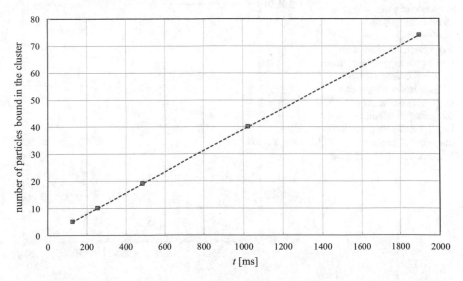

Fig. 4.8 Test of agreement of the real number of bound nanoparticles in five simulation phases with the required frequency (4.4) --.

Fig. 4.9 Simulation sequence of time evolution (for $t = 128$ ms) by configuration of the aggregated structure with corresponding physical times for the polydisperse system of spherical nanoparticles nC_{60} with statistical frequency distribution in Fig. 4.7 and with a mean diameter of 26 nm. Simulation cell (1×1 μm)

Fig. 4.10 Simulation sequence of time evolution (for $t = 256.4$ ms) by configuration of the aggregated structure with corresponding physical times for the polydisperse system of spherical nanoparticles nC_{60} with statistical frequency distribution in Fig. 4.7 and with a mean diameter of 26 nm. Simulation cell (1×1 μm)

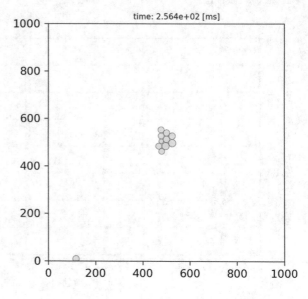

Fig. 4.11 Simulation sequence of time evolution (for $t = 487.2$ ms) by configuration of the aggregated structure with corresponding physical times for the polydisperse system of spherical nanoparticles nC_{60} with statistical frequency distribution in Fig. 4.7 and with a mean diameter of 26 nm. Simulation cell ($1 \times 1 \mu$m)

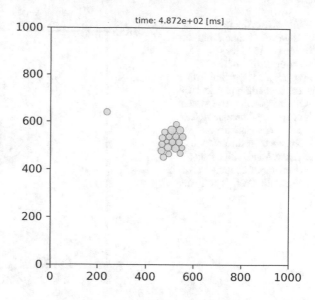

Fig. 4.12 Simulation sequence of time evolution (for $t = 1026$ ms) by configuration of the aggregated structure with corresponding physical times for the polydisperse system of spherical nanoparticles nC_{60} with statistical frequency distribution in Fig. 4.7 and with a mean diameter of 26 nm. Simulation cell ($1 \times 1 \mu$m)

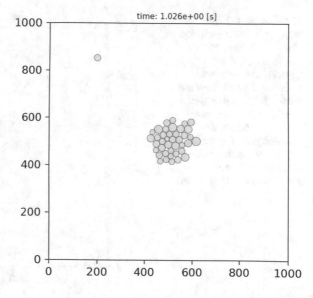

Fig. 4.13 Simulation sequence of time evolution (for $t = 1538$ ms) by configuration of the aggregated structure with corresponding physical times for the polydisperse system of spherical nanoparticles nC_{60} with statistical frequency distribution in Fig. 4.7 and with a mean diameter of 26 nm. Simulation cell (1×1 μm)

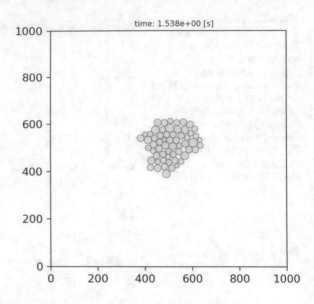

Fig. 4.14 Simulation sequence of time evolution (for $t = 1897$ ms) by configuration of the aggregated structure with corresponding physical times for the polydisperse system of spherical nanoparticles nC_{60} with statistical frequency distribution in Fig. 4.7 and with a mean diameter of 26 nm. Simulation cell (1×1 μm)

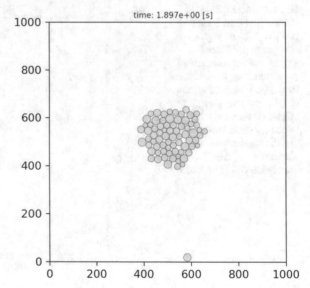

References

1. R. Dvorsky, J. Trojkova, J. Lunaček, K. Piksova, O. Černohorsky, Synthesis of inorganic nanofibers and lamellar structures with large specific surface by means of controlled vacuum freeze-drying process, in *NANOCON 2011*, pp. 58–63 (2011)

2. R. Dvorský, J. Lunacek, A. Sliva, K. Piksova, Formation of Lamellar and Fibrillar microaggregates of silicon nanopartocles during vacuum freeze drying of water suspension, in *5th International Conference on Surfaces, Coatings And Nanostructured Materials* (2010), pp. 612–614

3. R. Dvorsky, J. Trojková, J. Kukutschová, J. Luňáček, V. Vodárek, J. Holešinský, Synthesis and properties of lamellar aggregates of globular fullerene nano-aggregates. Adv. Sci. Eng. Med. **6**(4), 449–453 (2014). https://doi.org/10.1166/asem.2014.1519

4. R. Dvorský, Czech-303513 (2012), USA-US 9,410,739 B2 (2016), Japan-JP5961266 B2 (2016), China-CN 103842755 B (2016), Russia-RU 2599282 (2016), European Patent-EP2751508B1 (2017)

5. D. Krpelík, R. Dvorsky, P. Mančík, J. Bednář, L. Svoboda, Mathematical modelling of Lamellar aggregation of dispersed globular nanoparticles nC_{60} on the Interface Upon Sublimation of Water Molecules from Rapid Frozen Dispersion. J. Nanosci. Nanotechnol **19**(5), 2671–2677 (2019). https://doi.org/10.1166/jnn.2019.15856

6. J. Israelachvili, *Intermolecular and Surface Forces* (Elsevier, 2011)

7. H.-J. Butt, M. Kappl, *Surface and Interfacial Forces* (Wiley-VCH Verlag GmbH & Co. KGaA, Weinheim, 2010)

8. L. Bergström, Hamaker constants of inorganic materials. Adv. Colloid Interface Sci. **70**, 125–169 (1997). https://doi.org/10.1016/S0001-8686(97)00003-1

9. A. Patnaik, Structure and dynamics in self-organized C_{60} fullerenes. J. Nanosci. Nanotechnol. **7**(4), 1111–1150 (2007). https://doi.org/10.1166/jnn.2007.303

10. J. Hare, Properties of carbon and C60-data compliled by the CSC and the Sussex Fullerene Group. http://www.creative-science.org.uk/propc60.html

11. R. Dvorský, P. Praus, P. Mančík, J. Trojková, J. Lunacek, Preparation of globular nano-aggregates using microemulsion crystallization. NANOCON **2013**, 269–274 (2013)

12. T.N. Veziroglu, S.Y. Zaginaichenko, D.V. Schur, B. Baranowski, A.P. Shpak, V.V. Skorokhod (eds.), *Hydrogen Materials Science and Chemistry of Carbon Nanomaterials*, vol. 172 (Springer Netherlands, Dordrecht, 2005)

Chapter 5
Lamelary Aggregation of Nanoparticles from Frozen Liquid on the Sublimation Interface—Experimental Preparation and Application of Nanomaterials

5.1 Experimental Preparation and Characterization of Nanomaterials by the Method of Controlled Sublimation

5.1.1 Bottom-up Preparation of Globular Fullerite Nanoparticles nC_{60}

In our previous work [1], we described a new and simple method of microemulsion crystallization, which we used to prepare a spherical fullerite nanoparticles nC_{60}, see Fig. 5.1 (right), subsequently became the starting material for experiments of real self-organization in controlled sublimation.

Unlike standard two-component microemulsion methods, microemulsion crystallization works with only one component dissolved in "oil" microdroplets, which are dispersed by high-speed turbulent mixing in an intense ultrasonic field in an aqueous medium. Gradual evaporation of the solvent under the stated conditions and simultaneous bubbling with gas increases the concentration of fullerene molecules dissolved in the volume of the microdroplet (see the diagram in Fig. 5.1, left) until it is supersaturated.

In the supersaturated solution, the nC_{60} nanoparticles nucleate and gradually grow to the maximum final size, which contains all the C_{60} molecules originally dissolved in the volume of the microdroplet with radius R_0. The question arises as to whether more than one nucleus might nucleate and grow within a single microdroplet. Given that the probability of heterogeneous nucleation at the phase interface is many times higher in the processes of new phase formation than homogeneous nucleation, it is highly probable that after nucleation of the first nucleus its phase surface is absolutely dominant recipient of other C_{60} molecules from supersaturated solution. In

R. Dvorsky et al., *Nanoparticles' Preparation, Properties, Interactions and Self-Organization*, SpringerBriefs in Applied Sciences and Technology, https://doi.org/10.1007/978-3-030-89144-2_5

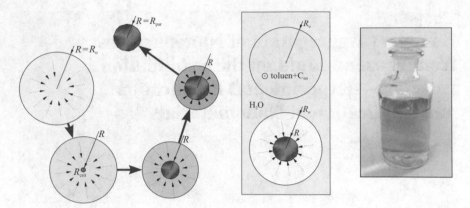

Fig. 5.1 Left: Development scheme of an individual microdroplet of a C_{60} molecules solution in toluene with an initial radius of R_0. During the evaporation of the solvent, the supersaturation and nucleation of the critical solid phase with radius R_{crit} gradually increase. The growth continues until the solvent is completely evaporated which creates nanoparticle of size R_{par}. Middle: Scheme of the initial and final state of the emulsion microreactor (droplet) in an aqueous medium as a basis for calculating the dependence of the final particle size on the concentration of C_{60} molecules in the toluene droplet. Right: Photograph of the final aqueous nanodispersion of spherical nanoparticles nC_{60} with mean size of 26 nm

this situation, further nucleation can occur with an extremely low probability, and we can make a realistic assumption: "1 droplet $(R_0) \rightarrow$ 1 nanoparticle (R)".

As a result, all the mass m of fullerene originally dissolved in the initial microemulsion droplet of radius R_0 accumulates in the final solid spherical nanoparticle of radius $R \rightarrow R_{par}$

$$c_0 \left(\frac{4}{3} \pi R_0^3 \right) = m = \rho_{sol} \left(\frac{4}{3} \pi R^3 \right) \rightarrow R_{par}(c_0, R_0) = \left(\frac{c_0}{\rho_{sol}} \right)^{\frac{1}{3}} R_0. \tag{5.1}$$

With a known density of the crystalline fullerite $\rho_{sol} = \rho_{nC_{60}} = 1750$ kg m^{-3}, we can prepare spherical fullerite nanoparticles of the desired radius R_{par} by choosing the mean size R_0 of the microdroplets and the volume concentration c_0 (kg m^{-3}) of the fullerite nanoparticle in solution. If we establish the parameter $C_{C_{60}}$ (%) as a percentage fraction of the saturated concentration of fullerene in toluene c_{sat} (kg m^{-3}), the (5.1) becomes

$$R_{par}\left(C_{C_{60}}, R_0\right) = \left(\frac{c_{sat}}{\rho_{sol}} \right)^{\frac{1}{3}} R_0 \left(\frac{C_{C_{60}}}{100} \right)^{\frac{1}{3}}. \tag{5.2}$$

Let R_{00} be the minimum radius of the emulsion droplet within the given conditions at the minimum volume fraction size of microdroplets $C_{dis} \approx 1\%$ and at constant dispergation intensity [ultrasound (W L^{-1}) and stirring (rpm)]. During emulsification, mechanical energy is transported from the surrounding aqueous medium to the microdroplets, which gradually creates a steady state with the same energy density in both liquid phases. In the first approximation, the hydrodynamic equilibrium of two immiscible phases (water–toluene) leads to the equilibrium radius R_0, which is a linear function of the volume fraction $R_0(C_{dis}) = R_{00} + \alpha C_{dis}$ with the proportionality coefficient α. The final radius R_{par} of nanoparticles is then a function of two variables according to formula (5.3)

$$R_{par}\left(C_{C_{60}}, C_{dis}\right) = \left(\frac{\rho_{sat}}{\rho_{sol}}\right)^{\frac{1}{3}} (R_{00} + \alpha C_{dis})\left(\frac{C_{C_{60}}}{100}\right)^{\frac{1}{3}}. \tag{5.3}$$

The linear dependence in the middle of the right side of Eq. (5.3) corresponds to the first hydrodynamic approximation estimate and was verified in the experiments of microemulsion preparation in the range $C_{dis} = (1\text{--}7)$ vol.%. For lower fractions below 1%, the results were already quite scattered and due to the very low efficiency of nanoparticle preparation at such low fractions, the applicability of Eq. (5.3) was limited from below by the lower limit of the mentioned interval (1–7) vol.%.

Good agreement of the regression model with the experimental data in Fig. 5.2 confirms its realistic character, suitable for practical application in the preparation of spherical nC_{60} nanoparticles of the required size.

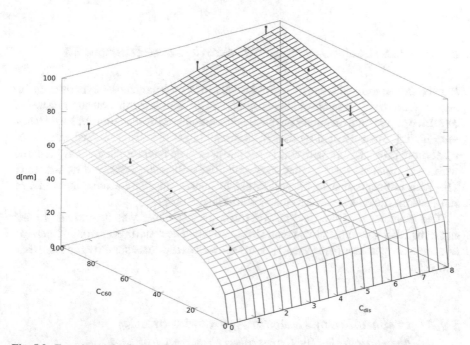

Fig. 5.2 Experimental values ■ of the nC_{60} nanoparticles diameter d were measured by the DLS dynamic light scattering method on a Malvern Zetasizer Nano ZS instrument. Data were interpolated by two-parameter nonlinear regression $d = 2R_{par}(C_{dis}, C_{C_{60}})$ according to Eq. (5.3). Regression deviations are illustrated by blue vertical lines ▮

Fig. 5.3 Micrograph of
fullerite nanoparticles
prepared by microemulsion
crystallization for controlled
sublimation experiments,
taken on a TEM JEOL 200
CX, 200 kV

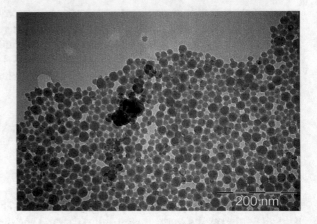

The micrograph in Fig. 5.3 shows nC_{60} nanoparticles, which were prepared by the above-mentioned microemulsion crystallization method and would be used in subsequent self-organization experiments at the sublimation interface. Their globular form confirms the expected spherical shape of the final particles, which also confirms the above assumption "1 droplet $(R_0) \rightarrow$ 1 nanoparticle (R)".

5.1.2   DLS Analysis of nC_{60} Nanoparticle Size Distribution

For the preparation of fullerite nanoparticles for self-organization experiments, an aqueous microemulsion of toluene droplets of C_{60} solution with a mean diameter of 382 nm was prepared by purely physical methods without the usual use of surfactants; see Fig. 5.4. Measurement by the DLS dynamic light scattering method was somewhat more difficult than with solid nanoparticles. The microemulsion was prepared with an optimal pH level so that the ζ-potential sufficiently stabilized its structure. The measurements had to be performed in quartz cuvettes because the walls of standard plastic cuvettes were etched with toluene drops.

The diameter d of the final nC_{60} fullerite nanoparticles was determined by the standard DLS method, and its mean size is 26 nm. The resulting aqueous dispersion of these nanoparticles (see Fig. 5.1, right) showed exceptional stability for more than one year.

5.1.3   Preparation of Lamellar Aggregates of nC_{60} Nanoparticles by Controlled Sublimation Method

Self-assembled aggregates were prepared from a frozen dispersion of nC_{60} nanoparticles with a mean size of 26 nm and a statistical distribution in Fig. 5.4 (left). During

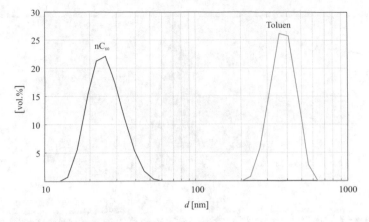

Fig. 5.4 Statistical distribution of the volume of dispersed fractions according to the diameter d of particles, measured by the DLS method on a Malvern Zetasizer Nano ZS instrument. Right: ——— droplets of a microemulsion of a solution of C_{60} in toluene with a mean size of 382 nm in diameter. Left: — nC_{60} nanoparticles with a mean diameter of 26 nm

controlled sublimation, the temperature at the sublimation interface was maintained at -40 °C and the pressure in the vacuum recipient at $p \approx 12.3$ Pa. This value is 96% of the saturated vapor pressure at a given temperature and also served as the basis for the physical analysis of the process as it was shown in previous Chap. 3.2.4. The controlled sublimation process was experimentally implemented in two variants using the same nanoparticle dispersion. In the first mode, the outlet surface of the chamber (flat dish) was covered with a fine plastic sieve in Fig. 5.3, the purpose of which was to reduce the pressure gradient above the sublimation interface and to minimize the destructive effect of the sublimation wind. In the second mode, sublimation took place from the chamber completely open to the vacuum recipient with the full intensity of the sublimation wind (Fig. 5.5).

In the first mode, when the sublimation wind was weakened, the statistical size distribution showed a relatively wide distribution, as shown in Fig. 4.7.

Fig. 5.5 Sublimation chamber covered with a mesh, partially suppressing the intensity of the sublimation wind (similar to the Knudsen cell)

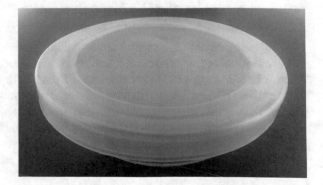

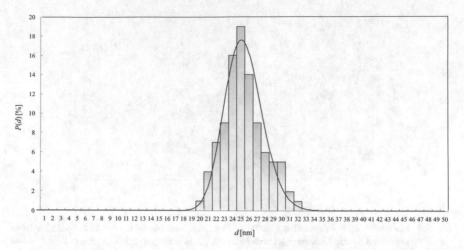

Fig. 5.6 Statistical distribution frequency function $P(d)$ of set nC_{60} nanoparticles according to their diameter d with normal regression —, obtained by image analysis of a representative selection of 150 nanoparticles of the TEM micrograph of a lamellar aggregate of fullerite nC_{60} with a mean size of 26 nm shown below in Fig. 5.8

   Image analysis of a representative selection of 240 nanoparticles of the TEM micrograph in Fig. 3.4 (left) then divided them into 50 size categories.

   The output material prepared in the second mode with full sublimation wind intensity showed a much narrower statistical distribution, as shown in Fig. 5.6.

5.2   Characterization of Nanomaterials Prepared by the Method of Controlled Sublimation and Their Applications

5.2.1   *Electron Microscopy of TEM and SEM Lamellar Aggregates of nC₆₀ Nanoparticles*

The following series of images contain micrographs of lamellar nanostructures prepared by the method of controlled sublimation from frozen aqueous nC_{60} nanoparticles. The shown fragments were selected from the dry form of the material. To prepare the TEM image, a small portion of it was very carefully dispersed in a drop of demineralized water over a formvar-coated copper microgrid. After about ten seconds, the excess liquid was aspirated through the filter paper and the sample was then dried for two minutes. This method guarantees good fixation of lamellar fragments on the lattice and at the same time does not cause their deagglomeration.

Fig. 5.7 Micrograph of a lamellar aggregate of fullerite nC_{60} nanoparticles with a mean size of 26 nm in diameter, taken on a TEM JEOL 200 CX, 200 kV

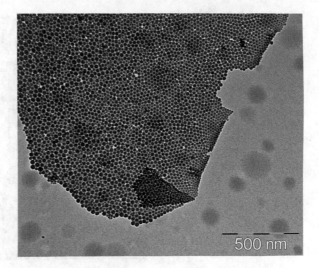

Figure 5.7 shows a higher frequency of significantly smaller particles. This material was prepared in the mode of attenuated effect of sublimation wind in a chamber covered by a sieve in Fig. 5.3.

Figure 5.68 shows a material prepared under the full action of a sublimation wind over an exposed surface. This statistical size distribution in Fig. 5.66 corresponds to the preparation in the mode without compensation of the destructive effect of the sublimation wind (Figs. 5.9, 5.10, 5.11 and 5.12).

The following images show a broader perspective view of the nC_{60} lamellar aggregates on a scanning electron microscope.

Fig. 5.8 Micrograph of a lamellar aggregate of fullerite nC_{60} nanoparticles with a mean size of 26 nm in diameter, taken on a TEM JEOL 200 CX, 200 kV

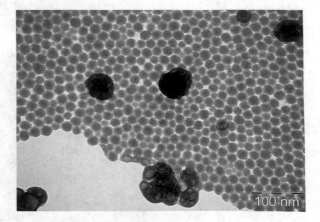

Fig. 5.9 Micrograph of a
lamellar aggregate of
fullerite nC$_{60}$ nanoparticles
with a mean size of 26 nm in
diameter, taken on a TEM
JEOL 200 CX, 200 kV. The
image shows cracks in the
lamellar structure, the
formation of which is caused
by both the pressure of the
sublimation wind and the
possible stress during later
manipulation

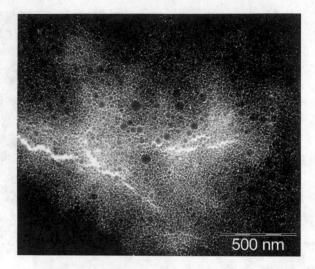

Fig. 5.10 Micrograph of a
lamellar aggregate of
fullerite nC$_{60}$ nanoparticles
with a mean size of 26 nm in
diameter, taken on a TEM
JEOL 200 CX, 200 kV. The
real image is overlaid on the
right by the final simulation
configuration at time 2.74 s
for a narrow statistical
distribution in Fig. 5.6, and a
high degree of similarity
between the two structures
can be noted

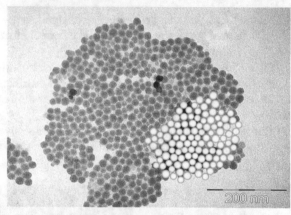

Fig. 5.11 Micrograph of a
lamellar aggregate of nC$_{60}$
fullerite nanoparticles with a
mean size of 26 nm in
diameter, taken on a
SEM-Phillips 515, 30 kV

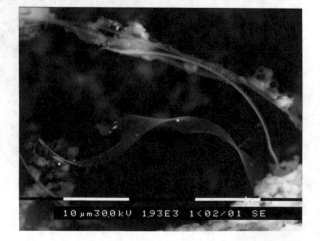

Fig. 5.12 Micrograph of a lamellar aggregate of fullerite nC_{60} nanoparticles with a mean diameter of 26 nm, taken on a SEM-JEOL JSM-7500F, 30 kV

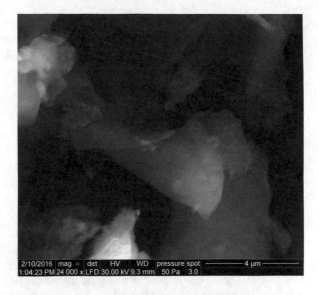

2/10/2016 mag □ det HV WD pressure spot — 4 μm —
1:04:23 PM 24 000 x LFD 30.00 kV 9.3 mm 50 Pa 3.0

5.2.2 Specific Surface Area of Lamellar Aggregates of nC_{60} Nanoparticles and Degree of Aggregation Shielding of the Surface by Individual Nanoparticles

From the statistical distribution P_v (d) of the total volume of nC_{60} nanoparticles from Fig. 4.7 shows, based on formula (3.23), the theoretical size of their unshielded specific surface area is 161 m^2 g^{-1}. During standard thermal evaporation of water molecules from the liquid state of the nanodispersion, the specific surface area has a very low value of 5.2 m^2 g^{-1}, which corresponds to about 3.3% of the theoretical size of the set of isolated particles.

Fig. 5.13 Photography of approx. 3 g of lamellar aggregates of nC_{60} nanoparticles ($\overline{d} = 26$ nm) in a sublimation vessel. The material was prepared by the method of controlled sublimation from an aqueous dispersion containing 3 wt% of nanoparticles of dispersion fraction A, its SSA is 145 m^2 g^{-1}

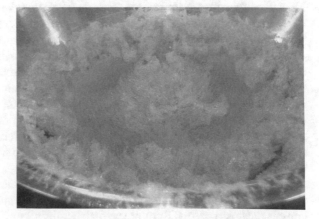

The experimental SSA value ($145 \text{ m}^2 \text{ g}^{-1}$) of material on Fig. 5.73 was determined by dynamic BET analysis. Compared to the theoretically calculated value, this indicates a very low degree of mutual shielding—about 10%.

5.2.3 Application of Nanomaterials Modified by the Method of Controlled Sublimation

5.2.3.1 Preparation of Efficient Photocatalytic Lamellar Nanostructures Core–Shell Si–ZnO with High Specific Surface Area

Environmental contamination by industrial pollutants, such as nitrogen oxides, is currently a highly topical issue. Long-term measurements of the concentration of nitrous oxide in the atmosphere show an annual increase of more than 0.2% [2]. Due to the destructive effects of nitrogen oxides on the ozone layer and their high stability, the problem of their degradation to harmless components such as molecular oxygen and nitrogen is very current. One promising way to solve this problem is the wide application of photocatalytic nanomaterials, such as traditional TiO_2 and ZnO nanoparticles.

With respect to Langmuir's law of effective surfaces [3], the high specific surface area and increased reactivity of nanoparticles [4] are important factors in their application in the processes of photocatalytic degradation of pollutants. Due to the difficult practical manipulation of very small volumes, it is often advantageous to deposit nanoparticles on larger carrier microparticles [5, 6]. At the same time, the morphology of the dry powder, which is very strongly dependent on the drying technology, is also very important for optimal photocatalytic efficiency.

In publication [7], a method of controlled sublimation preparation of photocatalytic lamellar nanostructures core–shell Si–ZnO is presented. For the most efficient use of the photocatalyst, the core–shell structure with a photoactive ZnO layer deposited on a large surface of the lamellar structure is advantageous; see Fig. 5.4.

In contrast to the classic material form of pure nanopowder, the prepared composite material shows very good macroscopic manipulability and at the same time high photocatalytic efficiency. In particular, the problem of mutual shielding of the active surfaces of nanoparticles during their aggregation due to Hamaker adhesion interactions is minimized. For large lamellar nanostructures, this effect is significantly limited by the low probability of achieving a parallel configuration of close reaction surfaces. High specific surface of the resulting nanomaterial increases photocatalytic efficiency and shows significantly better results than the standard Aeroxid® TiO_2 P25 (Evonik Degussa); see Fig. 5.5.

Of particular importance for the photocatalytic efficiencies of the material is the choice of the optimal mode of vacuum sublimation and subsequent annealing of lamellar nanostructures. Both processes significantly affect the morphology of the

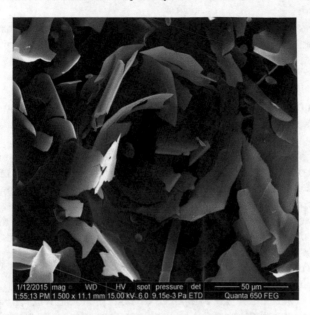

Fig. 5.14 SEM micrograph of lamellar aggregates of core–shell particles Si–ZnO with specific surface area 134 m^2 g^{-1}

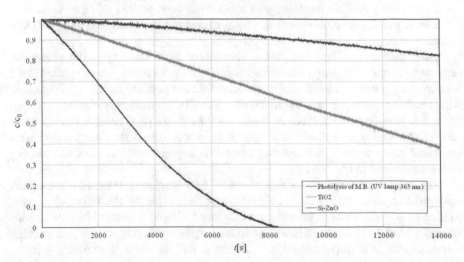

Fig. 5.15 Comparison of the time dependence of the relative concentration of the simulant (methylene blue) $c/c_0 = c(t)/c(0)$ during its photocatalytic degradation in the radiation field of a UV lamp (365 nm) with the TiO$_2$ standard Degussa P25 and Si–ZnO

final nanostructures. The above experimental results show that the new nanocomposite photocatalyst Si–ZnO is a very good candidate for application in many environmental and industrial processes.

5.2.3.2   Photocatalytic Reactivation of Silicate Nanosorbent Containing Graphitic Carbon Nitride (g-C$_3$N$_4$)

A significant problem in the application of sorption technologies is the costly disposal of saturated sorbent or its active recycling. It is active recycling to reuse sorption materials that is often a complex and expensive process for reactivating the sorption surface. A sorption nanostructure based on a silicate network, doped with Zn^{2+} ions, in which lamellar nanoparticles of g-C$_3$N$_4$ are dispersed, was created for the mentioned recycling applications. The high sorption capacity of the new nanomaterial is ensured mainly by the porous silicate nanostructure with the content of zinc oxide (ZnO-m·SiO$_2$). The reactivation effect is based on the simultaneous photocatalytic activity of zinc oxide in the UV region and g-C$_3$N$_4$ in the visible light region.

The highly porous silicate composite of both photocatalytic components was formed by the precipitation reaction of sodium water glass with modulus $m = 3$ in a zinc acetate solution in which lamellar g-C$_3$N$_4$ was dispersed. The precipitation reaction took place predominantly in a heterogeneous regime on their surface. It was on it that the insoluble ZnO-m · SiO$_2$ nanostructures nucleated and gradually cross-linked.

After the reaction was completed, subsequent rapid freezing of the composite reaction gel and application of controlled sublimation technology was applied and a composite nanomaterial with a high specific surface area of 366 m^2 g^{-1} was obtained. The resulting nanocomposite sorbent was subjected to a basic test of repeated sorption of a dye simulant (methylene blue) and its subsequent photocatalytic reactivation in various modes of artificial and natural light exposure. Experimental data with repeated reactivation in the presence of natural daylight without clouds in the sky are summarized in Fig. 5.6 [8, 9].

The achieved regeneration efficiency of the thus prepared photocatalytic sorbents showed a recovery rate of up to 94% of the original capacity. Photocatalytic degradation of sorbate when irradiated with ordinary daylight appears to be a good eco-friendly method of regeneration of saturated sorption structures and with a suitable air filter design with permanent light access it will be active in real time directly when the filter is applied.

5.2.3.3   High-Performance Sorptive-Photocatalytic Nanostructures on Fabrics

Controlled sublimation technology was applied in the fixation of photocatalytic nanoparticles based on zinc silicate by nanotextile support fibers. These regenerable materials should find significant application in the protection of persons against

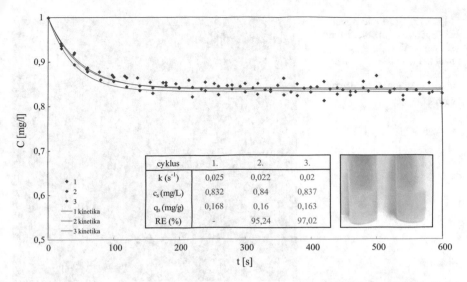

Fig. 5.16 Kinetic degradation curves of three sorption cycles of simulant (methylene blue) on ZnO-m · SiO₂/g-C₃N₄ composite material after repeated reactivation by photocatalytic degradation of sorbate in the sun. The table contains the kinetic characteristics of photocatalytic degradation: rate constants k, residual concentrations c_s of methylene blue after sorption, sorption capacity q_s and regeneration efficiency (RE). Photography of sorbent dispersion before (right) and after (left) after two hours of photocatalytic regeneration under daylight

contamination by harmful pollutants and CBRN substances [8]. The mentioned technology was used to fix in the structure of the nanotextile a photocatalytic layer with a high sorption surface ($410 \text{ m}^2 \text{ g}^{-1}$) [10]; see Figs. 5.7 and 5.8.

For most standard sorption fabrics, the preparation of powdered particles of sorption material takes place in separate processes, such as pulverization and carbon activation. Only then are these particles deposited and fixed in the supporting fibrous structures of the respective protective fabric.

Unlike most purely sorption materials such as activated carbon, the new nanomaterial was synthesized directly during the deposition of one of the reactants (Na_2SiO_3) into the space in the immediate vicinity of the nanofiber structures of the fabric. By modifying a previously verified synthetic method [7], a highly porous amorphous zinc silicate was prepared in the presence of graphene microparticles with a mean size of 564 nm directly in the structure of a nanofiber polyurethane fabric NnF MBRANE®-PUR 5gsm manufactured by PARDAM NANO4FIBERS Ltd.; see Fig. 5.19.

The purpose of doping the photocatalyst with graphene nanoparticles is to increase the mean lifetime of charges (e^-, h^+) by delocalizing the exciton after photon absorption. The scheme in Fig. 5.20 shows the mechanism of exciton charge decoupling by conducting electron in the graphene layer. This exciton dissociation significantly reduces the probability of recombination of both e^- and h^+ charges, increases their lifetime and the probability of diffusion to the reaction surface of the particles.

Fig. 5.17 SEM micrograph
of structure of photocatalytic
sorbent grains, fixed by
nanofibers NnF
MBRANE®-PUR 5gsm
(PARDAM NANO4FIBERS
s.r.o. (Ltd.))

Fig. 5.18 Macroscopic
photograph of nanofiltration
material under an optical
microscope Olympus BX 51

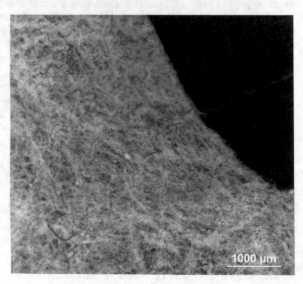

Mentioned effect of the longer presence of charges on the surface of the material
accelerates the kinetics of photocatalytic reactions and increases their degradation
efficiency.

At the surface, the charges then participate in the respective photocatalytic reactions. In the case of silicate material prepared by the method of controlled sublimation
[7], these are the surfaces of pores in a network nanostructure. Figure 5.21 shows a
solid dispersion of graphene microparticles in a transparent material of fixed grains.

sublimation wind

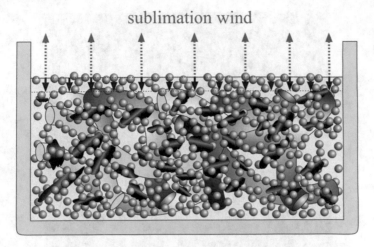

Fig. 5.19 Scheme of vacuum sublimation deposition of the microdispersion $Na_2SiO_3 + C_{graph}$ into the nanofibrous structure NnF MBRANE®-PUR 5gsm (*in unrealistic proportions*)

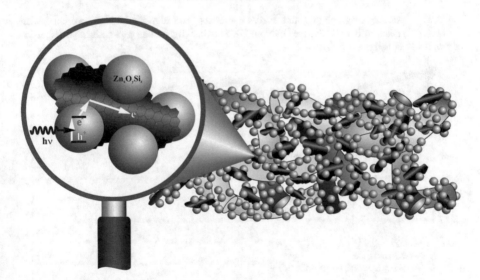

Fig. 5.20 Scheme of exciton creation and subsequent charge separation due to the presence of graphene lamellar particles in new fibrous material

These phenomena of exciton dissociation lead to a longer lifetime of surface charges and to an increase in the probability of their participation in photocatalytic reactions. There is thus a significant increase in photocatalytic efficiency compared to conventional standards, which is shown in Fig. 5.22.

In the photocatalytic degradation of methylene blue as a simulant, the half-life of the new carbon silicate material is about 2100 s, while the standard reference material

Fig. 5.21 Optical micrograph of grains of new photocatalytic material with high sorption surface $(410\,m^2\,g^{-1})$ taken with a microscope Olympus BX 51 in backlight mode with noticeable dispersion of grapheme lamellar microparticles

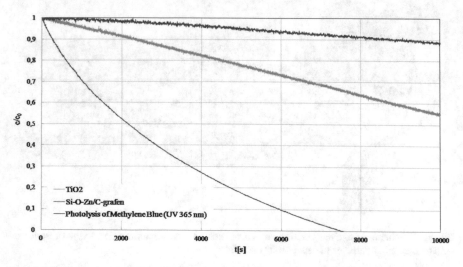

Fig. 5.22 Comparison of photocatalytic activity on decomposition of methylene blue by new material $(Si–O–Zn/C_{graph})$ with industry standard TiO_2 P25 Degussa under UV radiation with a wavelength 365 nm

TiO$_2$ P25 Degussa shows a value about five times higher than 11,000 s. Parameters of the photocatalytic sorption fabric treated in this way show very good grain fixation stability and high photocatalytic and sorption efficiency. In addition, the efficient photocatalytic decomposition of the adsorbed pollutants during the process prevents the negative effect of sorption saturation. These achieved results show a significant application potential of a new technology of deposition and fixation of photocatalytic sorbents in nanofiber structures.

5.2.3.4 Deposition of Sorption and Photocatalytic Material on Nanofibers and Textiles by Controlled Sublimation Method

The deposition of sorption material with photocatalytic regeneration directly on the surface of individual nanofibers is another possibility for the preparation of protective sorption textiles. The basic photocatalytic porous nanostructure of Si–O–Zn was in this case doped with dispersed nanoparticles of graphite or graphitic carbon nitride [11]. During the precipitation reaction, the insoluble silicate nanomaterial Si–O–Zn was fixed preferentially on the surface of the fibers by heterogeneous nucleation, and gradually coated most of their surface during the subsequent growth. Nanoparticles of graphite or g-C$_3$N$_4$ were also fixed in this basic sorption layer during the reaction. After rapid freezing, the fabric was subjected to a controlled vacuum sublimation process at temperature −41 °C. The resulting textile materials, see Figs. 5.23, and 5.24 showed specific surfaces of the order of hundreds of m^2 g^{-1}.

For most economically advantageous practical applications, the activation of photocatalytic regeneration by visible daylight is a natural condition. For these reasons, a very promising textile nanomaterial with g-C$_3$N$_4$ appears to be very promising (see Figs. 5.24 and 5.25).

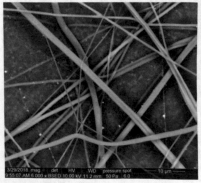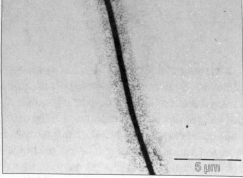

Fig. 5.23 Left: SEM micrograph of NnF MBRANE®-PUR 5gsm fibers from PARDAM and Right: detailed TEM micrograph of a fiber coated with a porous Si–O–Zn/C$_{graph}$ nanostructure with a specific surface area of 410 m^2 g^{-1}. Graphene nanoparticles are visible in the silicate coating of the fiber [11]

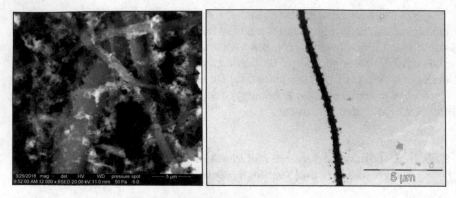

Fig. 5.24 Left: SEM micrograph and Right: detailed TEM micrograph of PARDAM NnF MBRANE®-PUR 5gsm fiber, coated with dispersed nanostructure Si–O–Zn/g-C$_3$N$_4$ with SSA ≈ 240 m^2 g^{-1} [11]

Fig. 5.25 Photograph of a fragment of the functional fabric NnF MBRANE®-PUR 5gsm by PARDAM with nanofibers coated with photocatalytic dispersion nanostructure Si–O–Zn/g-C$_3$N$_4$ [11]

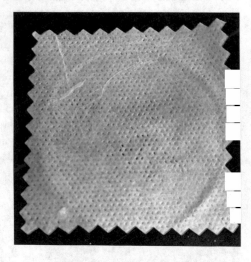

The semiconducting g-C$_3$N$_4$ has a band gap of about 2.7 eV [12]. This value corresponds to the maximum wavelength of excitation photons of about 450 nm, which already belongs to the blue part of visible light. It is the photocatalytic activity in visible light that is very important for the practical use of sorption fabrics with photocatalytic regeneration. The efficiency and rate of regeneration then determine whether the regeneration process will only slow down the achievement of sorption saturation or will function in real time as a continuous degradation of the adsorbed pollutants. Mentioned parameters were tested in the experimental apparatus (Fig. 5.26) in the degradation of indigo carmine dye (IC) as a simulant.

Degradation measurement was performed by a new in-situ method [9] by continuously measuring the decrease in optical absorption of the aqueous IC solution over the active fabric with a diameter of $d = 55$ mm at the bottom of the reactor in Fig. 5.26.

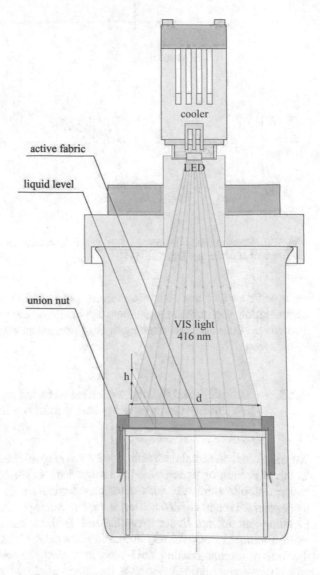

Fig. 5.26 Experimental photocatalytic reactor with depth $h = 2.5$ mm of photocatalytic bath for aqueous solution of simulant (MB—methylene blue) and diameter $d = 55$ mm of the tested fabric circular area [13, 14]

The decomposition was represented by the decolorization of the dye simulant solution when irradiated with blue visible light from a 10 W LED source at a wavelength of 416 nm. The measurement results in Fig. 5.27 confirm the very good efficiency of the photocatalytic regeneration of the sorption fabric with a half-life of the simulant of about 210 s.

In Fig. 5.27, it is a graph of photocatalytic degradation kinetics with a rate constant of 3.3 m s^{-1}. In addition, mechanical abrasion tests have shown very good fixation stability of the nanomaterial on the nanofibers, and the controlled sublimation method has proven to be an effective means of drying the active fabric with minimal loss

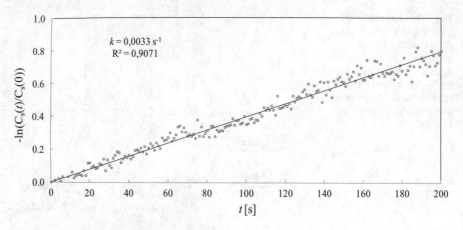

Fig. 5.27 Experimentally determined degradation kinetics of dye simulant indigo carmine (IC) over nanofiber fabric (see Fig. 5.16) [11]

of specific surface area. Sorption fabrics prepared in this way with photocatalytic regeneration will provide significant applications, especially in the production of protective equipment and clothing against permanent contamination by undesirable substances.

5.2.3.5 Antimicrobial Synergistic Effect Between Ag and ZnO in a Silicate Composite Ag–ZnO · mSiO$_2$ with a High Specific Surface Area

Antimicrobial materials, especially silver nanoparticles, are already widely used in the inhibition of undesirable microorganisms in the environment and in special medical applications. As with other nanoparticulate materials, it is advantageous to deposit silver nanoparticles on a carrier micropowder material that facilitates handling and dosing. In our previous work [15], the carrier material was a network nanostructure of ZnO · mSiO$_2$ similar to silica gel, which is doped with zinc atoms in the form of semiconducting ZnO. After saturating the porous material ZnO · mSiO$_2$ with an aqueous solution of AgNO$_3$, the nanoparticles of metallic silver Ag inside the porous silicate structure were prepared by photoreduction by the action of UV radiation at 200 nm for 100 min. Heterogeneous nucleation of Ag nanoparticles should generally occur uniformly on the large photoactive surface of the open pores of the carrier silicate ZnO · mSiO$_2$ without a significant effect of its chemical composition. In such a case, the antibacterial activity of both agents (Ag, ZnO) should show a common additive character with respect to both concentrations. However, the hypothesis that heterogeneous nucleation of Ag metal particles should occur preferentially in their immediate vicinity due to the photoactivity of ZnO semiconducting elements was tested in this work. As a result of this close binding, a synergistic effect should be observed in their antibacterial action.

The obtained nanocomposite material (Fig. 5.28) was tested for antimicrobial activity in both gram-positive and gram-negative bacterial strains and Candida albicans yeasts. The hypothesis of a possible synergistic effect of silver nanoparticles in close binding with zinc oxide in the Ag–ZnO · mSiO$_2$ composite was experimentally verified by measuring the minimum inhibitory concentration (MIC) for *Escherichia coli, Pseudomonas, Streptococcus* and *Staphylococcus* strains; see Table 5.1.

The evaluated data indicate a relatively significant synergistic effect between the antibacterial action of both components Ag and ZnO. By comparison, the differences indicated in the last column Table 5.1 as a saving.

The graph in Fig. 5.89 illustrates the dependence of the synergistic difference on the size of the zinc fraction v_{Zn} in the range from $c_0Ag = 0.003$ to $c_0Zn = 2.979$. While in the area of values below the intersection of the curves $v_{Zn} = 0.27$ the system shows reduced efficiency, in the area above the intersection the synergistic

Fig. 5.28 SEM micrograph of a silicate composite nanostructure Ag–ZnO · mSiO$_2$ with a specific surface area of about 250 m^2 g^{-1}, on which larger submicron aggregates of silver particles are clearly visible

Table 5.1 Calculated partial minimum inhibitory concentrations (MIC) for both active ingredients Ag and Zn allow to express the corresponding synergistic effect in the form of a percentage saving of the material

MIC (mg ml^{-1})	c_0Ag	c_0Zn	cAg	cZn	Saving (wt%)
E. coli	0.003	2.979	0.008	0.809	45.2
Pseudomonas	0.003	7.447	0.010	1.090	70.5
Streptococcus	0.102	5.957	0.016	1.652	44.9
Staphylococcus	0.102	5.957	0.016	1.652	44.9

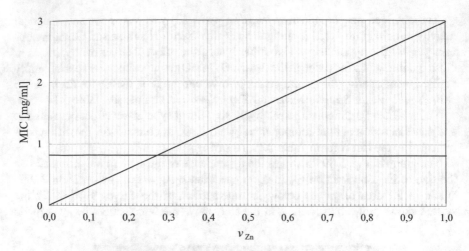

Fig. 5.29 Graph of the dependence — of the minimum inhibitory concentration for *E. coli* on the volume fraction of the v_{Zn} solution with MIC Zn (in the composite ZnO · mSiO$_2$ without Ag) with the complementary proportion of the solution with MIC Ag. Reference graph — of the constant value MIC (0.817), experimentally determined for the complex composite Ag–ZnO · mSiO$_2$

effect is significantly positive. These results proved the hypothesis of preferential heterogeneous nucleation of Ag nanoparticles in regions with increased content of the semiconducting component ZnO.

Fundamental antibacteriological tests for pure silver are usually performed in a molecular solution of silver nitrate, but for practical applications this source of silver ions is highly problematic. On the contrary, the nanocomposite powder material Ag–ZnO · mSiO$_2$ is neutral and easy to handle. In the process of controlled sublimation, its high specific surface area of 250 m^2 g^{-1} occupied by zinc ions with a close bond to silver metal nanoparticles was formed. As a possible explanation for the observed synergy, the formation of the Schottky semiconductor–metal transition can be considered, which represents a qualitatively new factor in the coexistence of both components. In our previous work [11], the possibility of deposition of this type of material on nanofibers and fabrics for antimicrobial applications in various types of filters was also experimentally verified.

References

1. R. Dvorský, P. Praus, P. Mančík, J. Trojková, J. Lunacek, Preparation of globular nano-aggregates using microemulsion crystallization, in *NANOCON 2013* (2013), pp. 269–274
2. J. Pérez-Ramírez, F. Kapteijn, K. Schöffel, J. Moulijn, Formation and control of N$_2$O in nitric acid production. Appl. Catal. B Environ. **44**(2), 117–151 (2003). https://doi.org/10.1016/S0926-3373(03)00026-2
3. H. Langmuir, the constitution and fundamental properties of solids and liquids. Part I. Solids. J. Am. Chem. Soc. **38**(11), 2221–2295 (1916). https://doi.org/10.1021/ja02268a002

4. B.J. Block, S.K. Das, M. Oettel, P. Virnau, K. Binder, Curvature dependence of surface free energy of liquid drops and bubbles: a simulation study. J. Chem. Phys. **133**(15), 154702 (2010). https://doi.org/10.1063/1.3493464

5. P. Praus, R. Dvorský, P. Pustková, O. Kozák, Precipitation of ZnS nanoparticles and their deposition on montmorillonite. Adv. Sci. Eng. Med. **3**(1), 113–118 (2011). https://doi.org/10.1166/asem.2011.1087

6. R. Dvorský, P. Praus, P. Mančík, J. Trojková, J. Lunacek, Preparation of globular nano-aggregates using microemulsion crystallization, in *NANOCON 2012* (2012), pp. 39–43

7. R. Dvorsky et al., Preparation of high-performance photocatalytic core-shell lamellar nanos-tructures ZnO-(Si)-ZnO With high specific surface area. Adv. Mater. Lett. **7**(9), 730–734 (2016). https://doi.org/10.5185/amlett.2016.6380

8. P. Mančík, J. Bednář, L. Svoboda, R. Dvorský, D. Matýsek, Photocatalytic reactivation of g-C_3N_4 based nanosorbent, in *NANOCON 2017* (2017), pp. 289–294 [Online]. Available: https://www.confer.cz/nanocon/2017/316-photocatalytic-reactivation-of-g-c3n4-based-nanosorbent

9. P. Mančík, Preparation and characterization of carbon-silicate nanosorbents with high specific surface area. VŠB - Technical University of Ostrava (2018)

10. R. Dvorský et al., Sublimation synthesis and deposition of high performance sorptive-photocatalytic nanostructures on fibres and fabrics, in *HAZMAT Protect 2016* (2016), pp. 40–46

11. R. Dvorský, L. Svoboda, J. Bednář, P. Mančík, D. Matýsek, M. Pomiklová, Deposition of sorption and photocatalytic material on nanofibers and fabric by controlled sublimation. Mater. Sci. Forum **936**, 63–67 (2018). https://doi.org/10.4028/www.scientific.net/MSF.936.63

12. Y. Kang, Y. Yang, L.-C. Yin, X. Kang, G. Liu, H.-M. Cheng, An amorphous carbon nitride photocatalyst with greatly extended visible-light-responsive range for photocatalytic hydrogen generation. Adv. Mater. **27**(31), 4572–4577 (2015). https://doi.org/10.1002/adma.201501939

13. L. Svoboda, N. Licciardello, R. Dvorský, J. Bednář, J. Henych, G. Cuniberti, Design and performance of novel self-cleaning g-C3N4/PMMA/PUR membranes. Polymers (Basel) **12**(4), 850 (2020). https://doi.org/10.3390/polym12040850

14. R. Dvorský, L. Svoboda, J. Bednář, P. Mančík, CZ2018421A3

15. J. Bednář et al., Antimicrobial synergistic effect between Ag and Zn in Ag-ZnO·mSiO$_2$ silicate composite with high specific surface area. Nanomaterials **9**(9), 1265 (2019). https://doi.org/10.3390/nano9091265

Conclusion

Self-organization of nanoparticles is currently a very important area in nanotechnology research. While bottom-up methods in current nanotechnology are applied to the composition of atoms, ions and molecules into nanostructures of very small dimensions, the problem of their interaction, especially in the field of nanoparticles, has so far been studied more on a theoretical level, as in the case of works of Landau and Hamaker. So far, the most extensive applications have reached theoretical outputs in describing the agglomeration of dispersed nanoparticles in liquid dispersions. The forces acting between particles in an aqueous medium are theoretically described in DLVO theory (*Derjaguin–Landau–Verwey–Overbeek*), which is based on the superposition of van der Waals forces and the repulsive action of the ζ-potential on the surface of particles. However, experiments with the self-organization of nanoparticles in a dry environment with significant application potential have not yet been carried out according to our own and international patent research.

One of the application possibilities is the preparation of lamellar nanoaggregates in the controlled sublimation of a frozen aqueous dispersion. Based on the study of the collective behavior of a set of a large number of nanoparticles in cold highly dilute gas (in the area of low vacuum pressures), a physicochemical analysis of the formation process of self-assembled nanoparticle aggregates was performed. The first result was the experience that lamellar structures of nanoparticles, bound by Hamaker interaction potential, are preferentially formed at the sublimation interface. The patent protection granted in the USA, Japan, Europe, China and Russia confirms the real novelty and topicality of physical research on this issue.

This work summarizes the current experience with the self-organization of nanoparticles released during vacuum sublimation on the surface of the dispersion of frozen fluids and brings the first theoretical and experimental results of the study of these phenomena. The theoretical core of the work consists in the physical analysis of self-organizing processes of globular nanoparticles nC_{60} at the sublimation interface and their mathematical modeling. The results of mathematical modeling showed that at real concentrations of nanoparticles in a frozen aqueous dispersion and a relatively

R. Dvorsky et al., *Nanoparticles' Preparation, Properties, Interactions and Self-Organization*, SpringerBriefs in Applied Sciences and Technology, https://doi.org/10.1007/978-3-030-89144-2

low vacuum, the effect of Brownian motion in dilute water vapor above the sublimation interface is negligible. In the first approximation, the simulation experiments were limited to 2D geometry above the phase interface, where the particles move in the water vapor and have only two translational degrees of freedom. The results of mathematical modeling showed a good agreement of the morphology of real lamellar aggregates with the simulation results. In this context, a specific tendency of nanoparticles with a narrower statistical size distribution to form denser and more regular lamellar nanostructures was also shown. A very interesting problem in this context appears to be the process of self-organization of monodisperse nanoparticles, which promises highly regular nanostructures with very promising new physical properties.

It is also necessary to draw attention to the already very useful applications of these materials in the field of sorption and catalysis, which have already been verified in several of our cited publications. In addition to high efficiency in the mentioned processes, a new class of sorption nanomaterials with a high specific surface area and permanent photocatalytic reactivation is particularly important. Recent experiences also show a very useful application of said nanomaterials in the form of air filter fabrics with permanent deactivation of trapped infectious agents. This effect is achieved by the action of electric charges emitted during the photocatalytic process on the surface of the filter in a radiant field of visible light.

Printed in the United States
by Baker & Taylor Publisher Services